NEAREST STA

How did the Sun evolve, and what will it bec
of its light and heat? How does solar activity affect the atmospheric conditions that make life on Earth possible? These are the questions at the heart of solar physics, and at the core of this book.

The Sun is the only star near enough to study in sufficient detail to provide rigorous tests of our theories and to help us understand the more distant and exotic objects throughout the cosmos. Having observed the Sun using both ground-based and spaceborne instruments, the authors bring their extensive personal experience to this story revealing what we have discovered about phenomena from eclipses to neutrinos, space weather, and global warming.

This second edition is updated throughout, and features results from the current spacecraft that are aloft, especially NASA's Solar Dynamics Observatory, for which one of the authors designed some of the telescopes.

Leon Golub is a Senior Astrophysicist at the Smithsonian Astrophysical Observatory, and has been studying the Sun and solar-type stars since the Skylab missions in 1973–74 and the Einstein Observatory in 1978. He is the head of the Solar-Stellar X-ray Group at the Harvard-Smithsonian Center for Astrophysics and has been involved in building and flying cutting-edge space instrumentation for the past thirty years. He is Chair of the Solar Physics Division of the American Astronomical Society and has written many popular articles on subjects ranging from astronomy and philosophy to music criticism.

Jay M. Pasachoff is the Field Memorial Professor of Astronomy at Williams College. He is a veteran of 58 solar eclipse expeditions, which have taken him all over the world to study the Sun over about five sunspot cycles. He received the Education Prize from the American Astronomical Society and the Jules Janssen Prize from the Société Astronomique de France; he is an honorary member of the Royal Astronomical Society of Canada. His undergraduate textbooks in astronomy, most recently the fourth edition of *The Cosmos: Astronomy in the New Millennium*, have been widely used. He is already involved in planning for education and public outreach for the 2017 total solar eclipse for which totality will stretch from Oregon to South Carolina and for which the whole of the continental United States and Canada will see at least a partial eclipse.

NEAREST STAR

The Surprising Science of Our Sun

Second Edition

Leon Golub
Harvard-Smithsonian Center for Astrophysics

Jay M. Pasachoff
Williams College

CAMBRIDGE UNIVERSITY PRESS

CAMBRIDGE
UNIVERSITY PRESS

32 Avenue of the Americas, New York, NY 10013-2473, USA

Cambridge University Press is part of the University of Cambridge.

It furthers the University's mission by disseminating knowledge in the pursuit of education, learning, and research at the highest international levels of excellence.

www.cambridge.org
Information on this title: www.cambridge.org/9781107672642

First edition published by Harvard University Press 2001
Second edition published by Cambridge University Press 2014

Printed in the United States of America

A catalog record for this publication is available from the British Library.

Library of Congress Cataloging in Publication Data
Golub, L. (Leon), author.
Nearest star : the surprising science of our sun / Leon Golub, Harvard-Smithsonian Center for Astrophysics, Cambridge, Massachusetts and Jay M. Pasachoff, Williams College, Williamstown, Massachusetts. – Second edition.
pages cm
Includes bibliographical references and index.
ISBN 978-1-107-67264-2 (pbk)
1. Sun. I. Pasachoff, Jay M., author. II. Title.
QB521.G65 2014
523.7–dc23 2013030423

ISBN 978-1-107-05265-9 Hardback
ISBN 978-1-107-67264-2 Paperback

To the future generations for whom we hold the Earth in trust.

Contents

Preface

Our Sun is a fairly ordinary star, a bit brighter than most but not exceptionally so. There are many stars much bigger and brighter, while most stars are smaller and fainter. The Sun is not an especially variable or active star, and it has no enormous chemical or magnetic peculiarities. It is not a very young star, nor is it old and nearing the end of its life. It is, in short, truly exceptional in only one way: it is very close to the Earth – in fact, at just the right distance to make life as we know it possible.

Most of us do not worship the Sun as did many in ancient civilizations, but we certainly should not take for granted the light and heat that it provides. Left to itself, the Earth would be a fantastically frigid rock at near absolute-zero temperature. If the Sun had been slightly more massive, its high temperature would have made the Earth's surface hot enough to melt lead. A smaller Sun would have left the Earth unbearably cold and possibly subject to high levels of radiation, since smaller stars tend to have higher levels of activity, giving off devastating ultraviolet and x-rays. Distance also matters. Had the Earth been closer, we might be as infernally hot as Venus; farther away and we might have been as cold and arid as Mars. We are in the position of Goldilocks, living at just the right distance from a just-right star.

Does this mean that the planet Earth is unique and that we live in a providential "best of all possible worlds"? There are dangers with this way of thinking, flattering as it is to human sensibilities, because it may foster a certain complacency, a feeling that things could not be otherwise. Since indeed other planets in our Solar System do not so far appear to support life, this implies that life requires some fairly unlikely conditions in order to flourish. On the other hand, granted that the probability of finding Earth-like conditions is small, the number of planets in the Universe is very large (probably billions in our Galaxy alone). This obviously increases the statistical likelihood of habitable planets. On this view, the Earth is not so much providentially unique as merely *rare*.

This in turn implies certain responsibilities for its inhabitants. Since life as we know it appears to be possible within only a narrow range of conditions, it would be prudent to know as much as we can about the star that provides the bedrock conditions on which our existence is founded. Moreover our newfound ability to alter the Earth's state on a global scale brings this need into sharp focus. For example, it is not enough for the Earth to be at the right distance from the Sun, and reflect back the right percentage of the solar light it receives. The Earth's atmosphere is also of major importance in determining the global temperature. Without it, the Earth would be colder by about 33°C (roughly 60°F), and therefore a frozen lump of ice. Right now, we are making small but significant changes to the composition of our atmosphere that are beginning to be large enough to produce major unpleasant effects. Do the natural variations in the Sun's brightness enhance or diminish these man-made effects? How do changes in solar activity affect the formation of ozone and atmospheric circulation and weather patterns?

This book explores the Sun in a comprehensive way for the non-scientific reader who wants to gain a general idea of the range and significance of solar physics. We explain what is known about the Sun and how this knowledge is acquired, discuss the origin of the Sun's light and heat, and explore how the Sun evolved and what it will become. We pay special attention to cutting-edge research on the Sun's outer atmosphere – the part that we can see – and the effects of this atmosphere on the Earth and the space around Earth. Unlike other stars, which are mere points in the sky, the Sun is so close that we can see its surface. We see sunspots form and gigantic explosive events erupt out toward the Earth. Thanks to careful measurements of the Sun's surface motions, we have recently even learned to "see" inside the Sun.

A quick tour of the contents of this book may convey our intention of being both informative and analytical:

Chapter 1. The Sun and its satellites: basic facts about the Sun and a few insignificant specks in its neighborhood.
Chapter 2. From gas to light: how a cloud of gas and dust turns into a 380,000,000,000,000,000,000-megawatt furnace.
Chapter 3. What we see: what our eyes, with a bit of help, tell us about the Sun.
Chapter 4. What we don't see: the colors we can't see and how we look inside the Sun.
Chapter 5. Eclipses: studying the Sun by covering it up.
Chapter 6. Space missions: what it's like beyond our atmosphere.
Chapter 7. The long haul: climate and the Sun.
Chapter 8. Space weather: the dangers lurking above our thin atmosphere.

The chapters may be read in any order, as they are largely self-contained. The first three chapters are similar to what is found in a standard astronomy book, although up to date and with emphasis on the subjects to be discussed in the rest of the book. Chapters 4–8 make up the main body, starting with our new ability to study the previously invisible parts of the Sun, followed by discussions of eclipse expeditions and space research from a personal perspective, as experienced by the authors. Chapter 7 deals with the relationship between the Sun and the Earth's climate, including the difficult issue of global warming. The last chapter concerns the shorter-term variations in the space around the Earth, which are now being called "space weather," and the reasons why we should be concerned about it.

Our book continues a fine tradition of descriptive books about the Sun for general audiences. We are proud to be following in the footsteps of Donald H. Menzel's *Our Sun*, with its first edition in 1949 and its second edition in 1959. We are also proud to be in the tradition of Robert W. Noyes's *The Sun, Our Star* (1982). There is much that is new on and under the Sun in the intervening years, and it is a pleasure to be able to describe it here. One of us (JMP) got his start in solar astronomy from both the distinguished scientists who were just listed. Donald Menzel took him, as a Harvard first-year student, to a total solar eclipse, which they saw from an airplane over the Massachusetts coast, and introduced him to the changing solar surface as part of a freshman seminar. Robert Noyes took him, as a graduate student, to his first professional observing experiences by inviting him to spend a summer working with him and with Jacques M. Beckers at the Sacramento Peak Observatory, Sunspot, New Mexico. That work developed into his thesis on the solar chromosphere, with Noyes as advisor. Subsequently, as the Donald H. Menzel Postdoctoral Fellow at the Harvard College Observatory, he worked with Professor Menzel in running a Harvard-Smithsonian expedition to the 1970 total solar eclipse in Mexico. He also collaborated with Dr. Menzel on eclipse expeditions to Prince Edward Island, Canada, in 1972, and to Kenya in 1973. It could not have been foreseen that Dr. Menzel's Harvard freshman seminar would begin JMP on his current set of 58 solar eclipse expeditions that have taken him around the world.

This book attempts to render all technical material in ordinary English. In other words, our text contains few, if any, equations. The book closest to the present volume is Ken Lang's excellent *Sun, Earth and Sky*. The main difference between the two is that we approach the story from a different point of view. Rather than present science as a series of prestigious accomplishments, we invite the reader into an open-ended process of discovery. We try to show what motivates the questions that are being framed in solar physics, and how instrumental developments and theoretical creativity work together in a dynamic way to gain better insight into the Sun. Our aim is to introduce a wide and diverse audience to the substance and importance of solar physics without straining the reader's patience. If we succeed in doing this, our efforts will be amply rewarded.

Although we primarily address non-scientists, we hope that technophiles may also find the discussions worthwhile, as we devote considerable attention to instrumentation. For those who want to pursue some of our topics in a more technical fashion and who have access to the World Wide Web or the Internet, the following are Web sites specializing in solar or solar-terrestrial matters:

- The Solar Data Analysis Center at NASA:
  http://umbra.nascom.nasa.gov/images/latest.html
- Solar Monitor:
  http://www.solarmonitor.org
- The Space Weather Prediction Center:
  http://www.swpc.noaa.gov

- The Sun Today:
  http://sdowww.lmsal.com
- National Solar Observatory:
  http://www.nso.edu
- Today's Space Weather:
  http://www.spaceweather.com
- Solar eclipse Web site from JMP for the International Astronomical Union:
  http://www.eclipses.info
- Eclipse maps:
  http://sites.williams.edu/iau-eclipses/reference-materials/#maps

Finally, despite our best efforts, there will inevitably be typos and errors. We apologize for this in advance, and plan to maintain an errata page with an updated list of corrections, located at:

http://www.williams.edu/Astronomy/sun.html

We encourage readers to send us typographical or other errors they find at:
lgolub@cfa.harvard.edu or eclipse@williams.edu.

Acknowledgments

LG would like to acknowledge helpful discussions with many friends and colleagues, especially: Spiro Antiochos, Guillaume Aulanier, Jay Bookbinder, Ed DeLuca, Eric Priest, Robert Rosner, and Harry Warren. Much of the work described in this book was supported by the National Aeronautics and Space Administration (NASA), and LG thanks Dave Bohlin, Bill Wagner, and George Withbroe for their help and support. He is especially grateful to his wife, Anne Davenport, for helping to make this book much better than it otherwise would have been. He welcomes the support and delight provided by family members and grandchildren Jessica/Casey/Ansel, Pablo/Liz, Manuel, and Charles/Jessica/Jacob.

JMP would like to thank the students at Williams College and former students who read portions of the manuscript and provided helpful comments: Dan Seaton, Misa Cowee, Alexandru Ene, Sara Kate May, Mark Kirby, and Kevin Russell. We thank Geoff Reeves for material about the Van Allen belts. Jay M. Pasachoff's work on solar astronomy in recent years has been supported by the National Science Foundation, by NASA, and by the Committee for Research and Exploration of the National Geographic Society. He thanks his wife, Naomi, for her collaboration in academic and nonacademic ways. He welcomes the family presence and support of children and grandchildren Eloise/Tom/Sam/Jessica and Deborah/Ian/Lily/Jacob.

We want to celebrate the memory of Donald Menzel and of Florence Menzel and to thank Bob and Harriet Noyes and Eugene Avrett for their long-term friendships. We both thank our editor of the first American edition, Michael Fisher of Harvard University Press, and the two (anonymous) reviewers selected by HUP, who provided many good suggestions for improving this book.

Many friends and colleagues have contributed by reading portions of the manuscript, by answering questions and by providing material. We especially thank: Hoyt Rogers, Becca McMullen, and Michele Sprengnether for constructive criticisms of several chapters. These friends have all provided valuable suggestions and advice, but of course any remaining errors in the text are entirely our responsibility. As noted at the end of the Preface, an updated list of errata may be found at the stated URL. We give special thanks to William Sheehan for numerous comments and suggestions.

We thank Madeline Kennedy for her excellent work on various aspects related to the book and its publication.

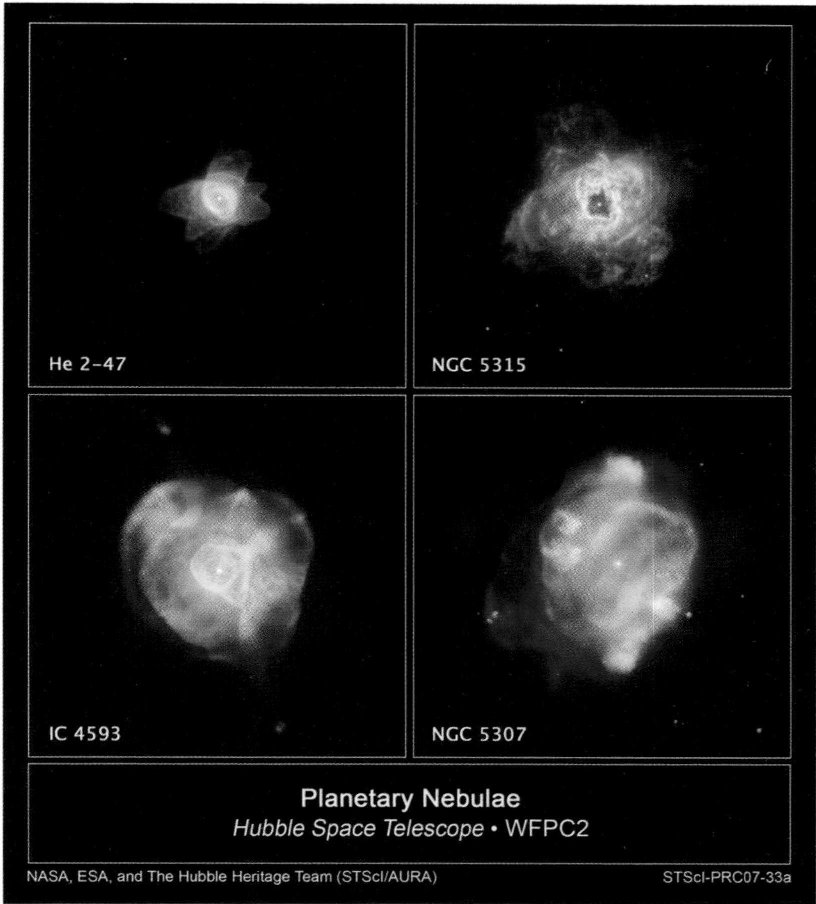

PLATE 1. An assortment of planetary nebulae, a phase of stellar evolution toward which the Sun is headed, as imaged by NASA's Hubble Space Telescope.

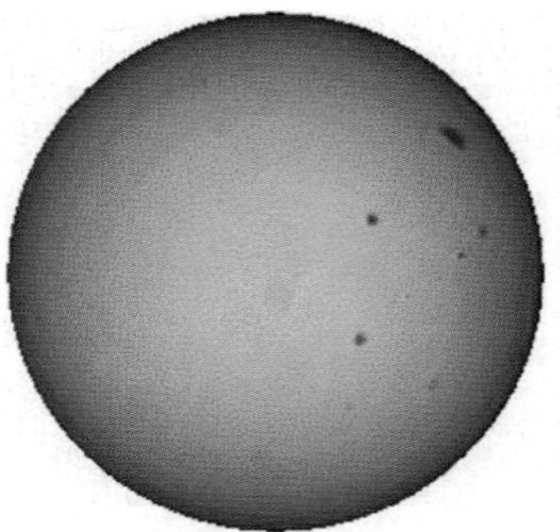

Intensity

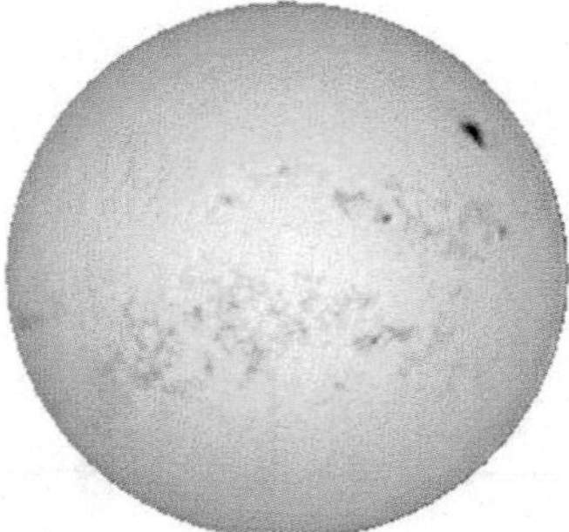

Modulation

Velocity

PLATE II. The intensity, corrected intensity (modulation), and velocity images of the solar surface from the Global Oscillations Network Group (GONG).

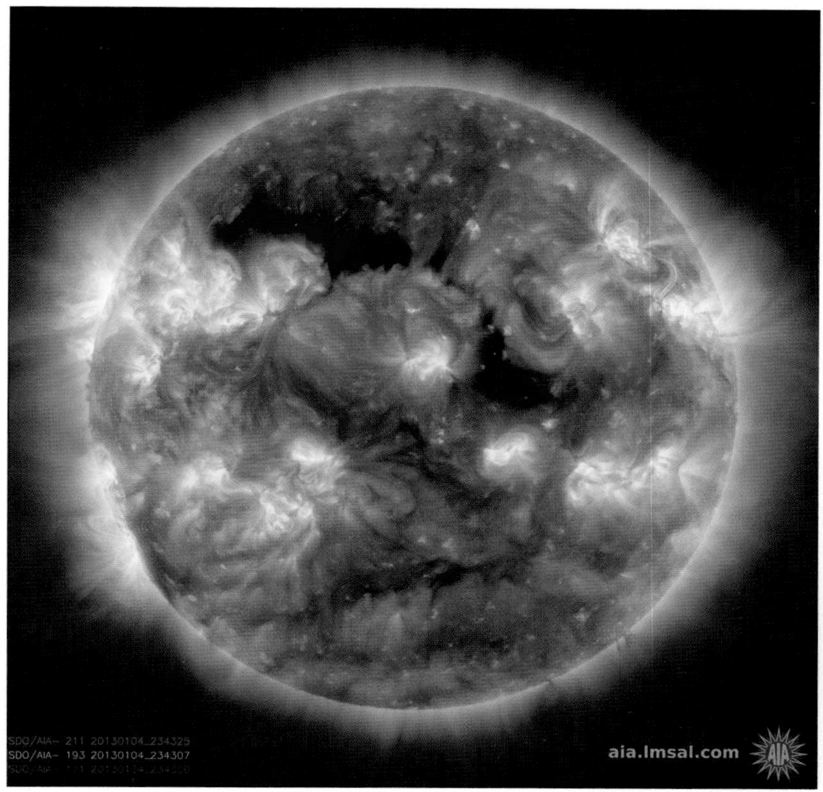

PLATE III. A view of the active high-temperature corona, observed by the Atmospheric Imaging Assembly on NASA's Solar Dynamics Observatory, January 3, 2013. The colorized image gives an approximate representation of the temperatures found in the corona, from below 1 million kelvin in dark blue to above 2.5 million kelvin in white.

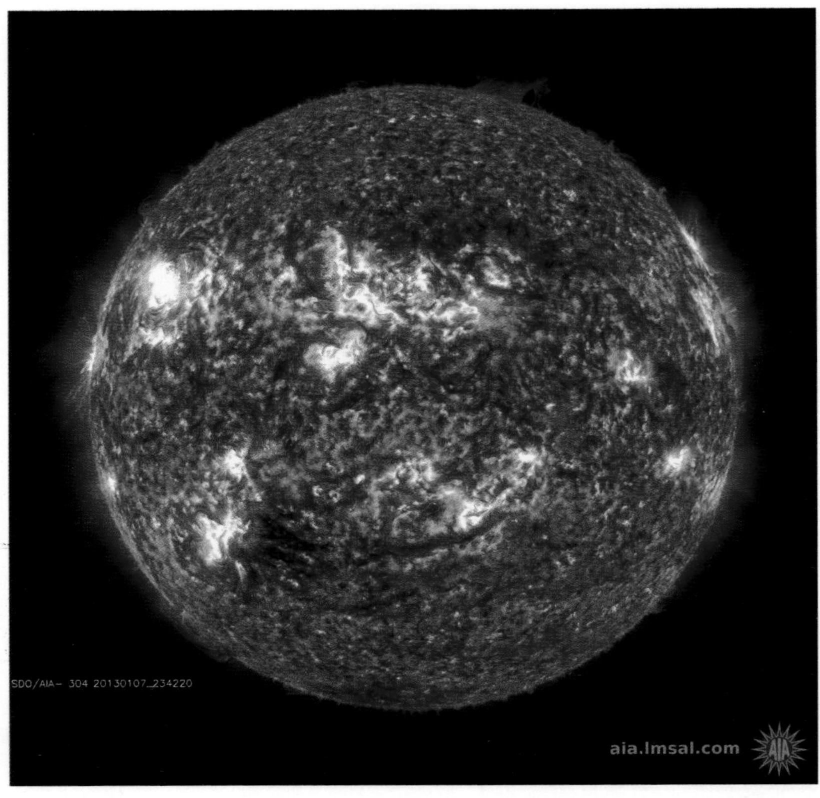

PLATE IV. A full disk image of the Sun in the light of He II at 30.4 nm on 7 January 2013, showing the portion of the solar atmosphere at temperatures intermediate between those of the photosphere and the corona.

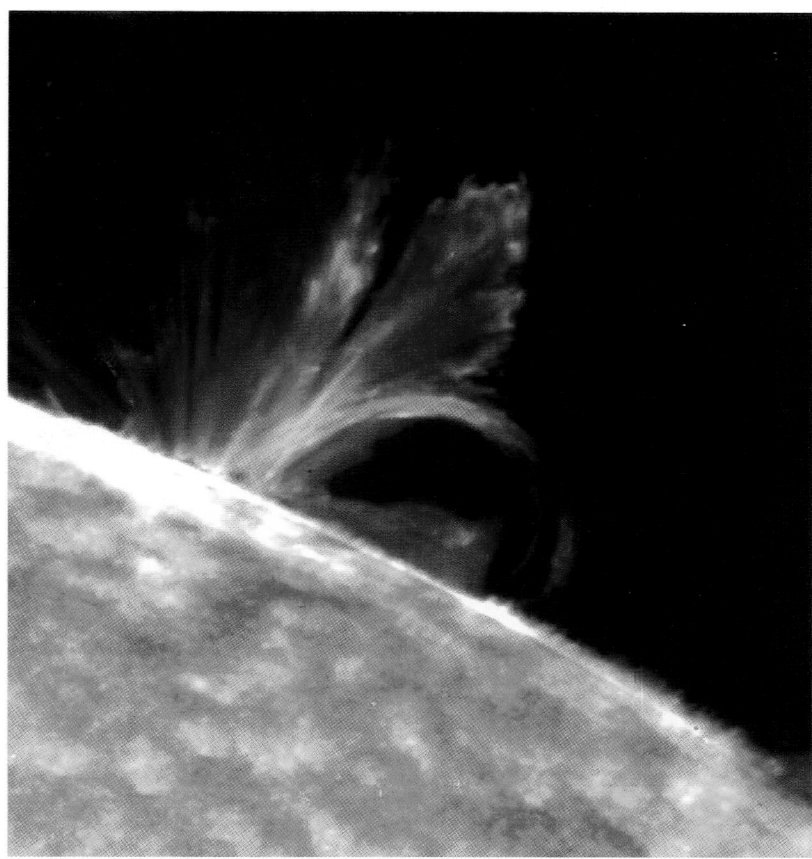

PLATE V. An image of chromospheric loops taken in the Lyman alpha line of hydrogen at 121.6 nm wavelength, May 19, 1998, from the Transition Region and Coronal Explorer (TRACE) satellite.

PLATE VI. A composite image of the corona at the time of the 2012 total solar eclipse taken from Australia, which is shown in the doughnut surrounding a Solar Dynamics Observatory/Atmospheric Imaging Assembly view of million-degree gas pasted over the dark silhouette of the Moon, with both surrounded by the view from the C2 coronagraph of the Solar and Heliospheric Observatory's Large Angle Spectrometric Coronagraph (SOHO/LASCO). One can trace many of the features from their origins on the side of the solar disk that is facing us through the inner and middle corona that is seen best at eclipses to the outer corona seen from SOHO.

PLATE VII. A view of the fine structure in a coronal active region at 1 million kelvin, as seen by the TRACE satellite on May 19, 1998. The AIA imagers on SDO have similar spatial resolution, with additional wavelength coverage and full Sun field of view.

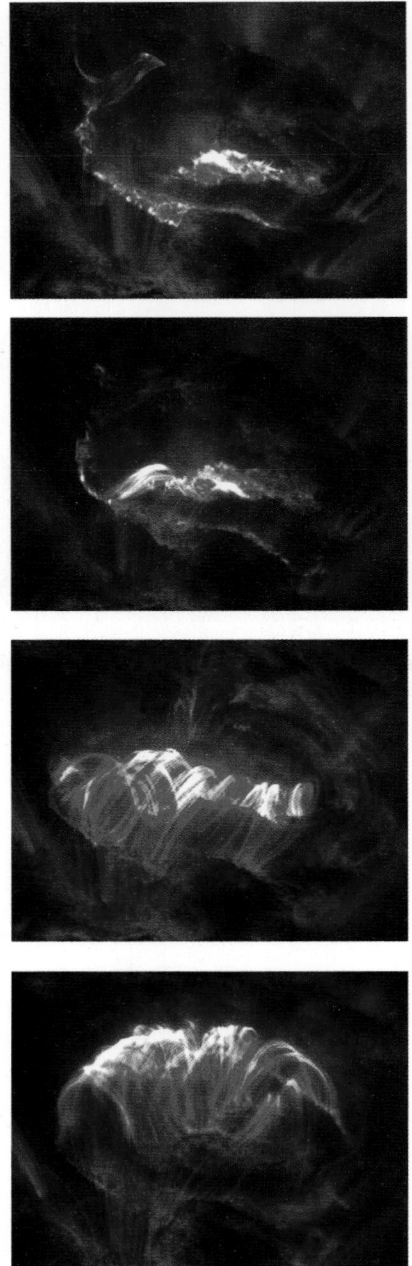

PLATE VIII. The progression of a large solar flare, from low-lying ribbons brightening, to the formation of postflare loops overlying and connecting the ribbons, as seen by the TRACE satellite.

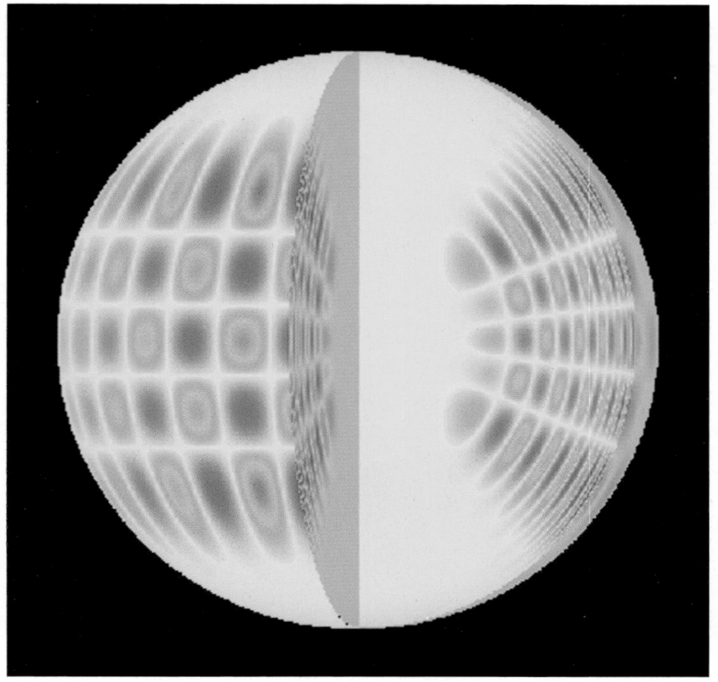

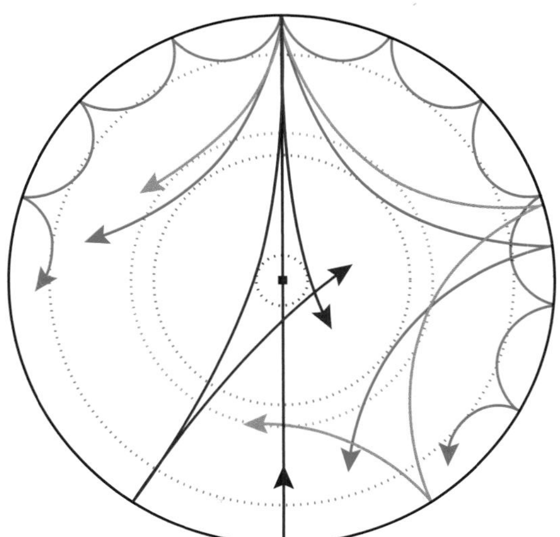

PLATE IX. (a) A mathematical model representing one of the many modes of oscillation of the Sun. (b) Waves travelling into the Sun are bent back toward the surface, where they can be studied to reveal properties of the interior.

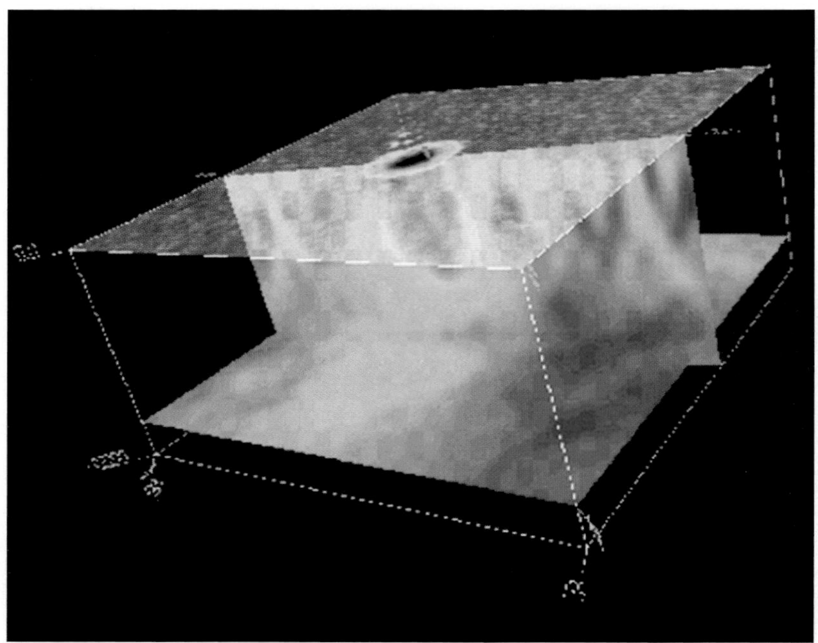

PLATE X. The surface and subsurface presence of a sunspot is revealed from detailed analysis of helioseismic data taken by the Michelson Doppler Interferometer (MDI) experiment on the Solar and Heliospheric Observatory (SOHO), June 18, 1998.

PLATE XI. NASA's STEREO Mission places two spacecrafts ahead and behind the Earth in its orbit, to provide stereoscopic views of solar eruptions as they leave the Sun and head toward the Earth.

PLATE XII. The Hinode spacecraft is shown during ground testing in Japan. The large cylindrical central tube is the optical telescope and the smaller black cylinder mounted on the right side is the X-ray Telescope (XRT). The solar panels are folded up along the left side of the optical telescope in this image.

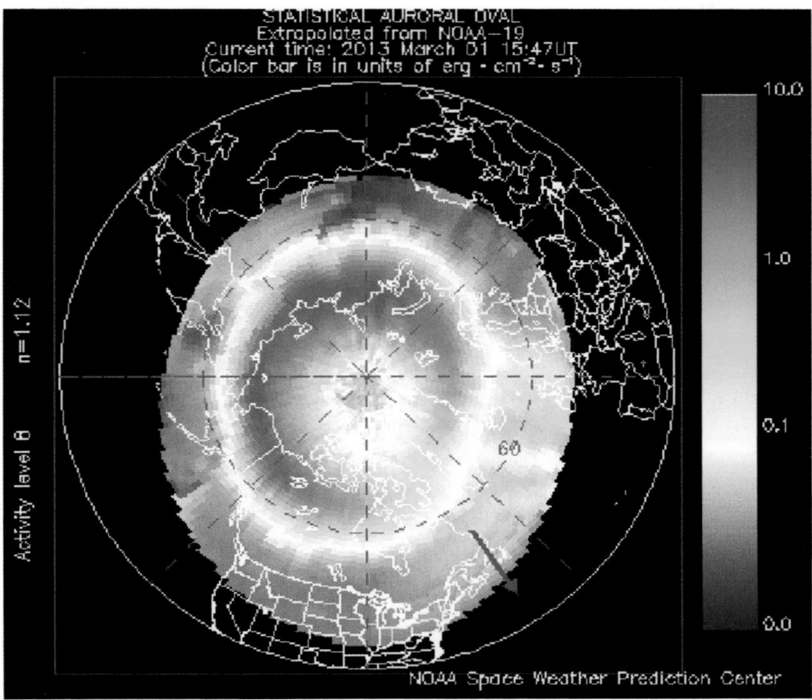

PLATE XIII. The auroral oval, as seen from space by NOAA's POES (Polar-orbiting Operational Environmental Satellite) spacecraft on March 1, 2013. Because of the symmetry in the Earth's dipolar magnetic field, there are typically auroras on both North (this page) and South (next page) magnetic poles simultaneously.

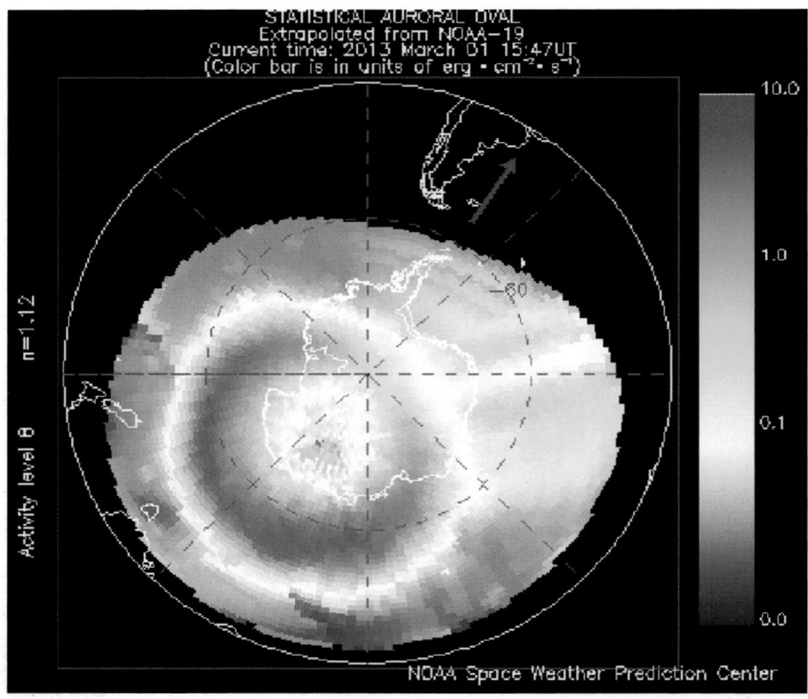

PLATE XIII. (*cont.*)

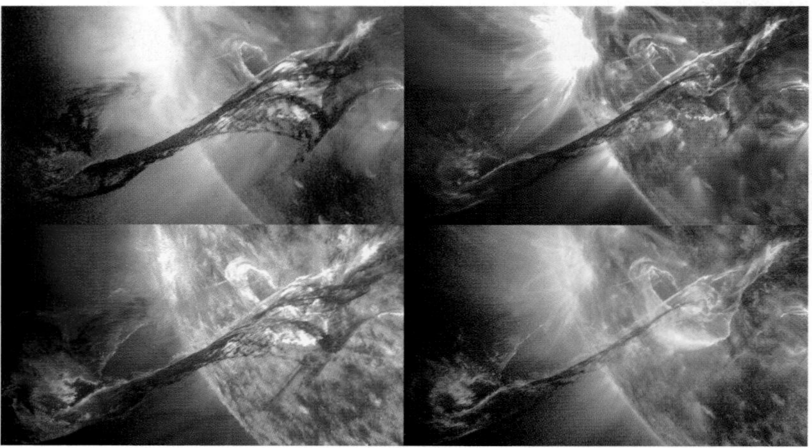

PLATE XIV. A view of a coronal mass ejection in four EUV wavelengths from the Atmospheric Imaging Assembly on NASA's Solar Dynamics Observatory. These images, taken at temperatures of 3 million kelvin (MK), 1 MK, 70 thousand K and 9 MK, show a large prominence in the process of being ejected on August 31, 2012, carrying most of the mass of the CME.

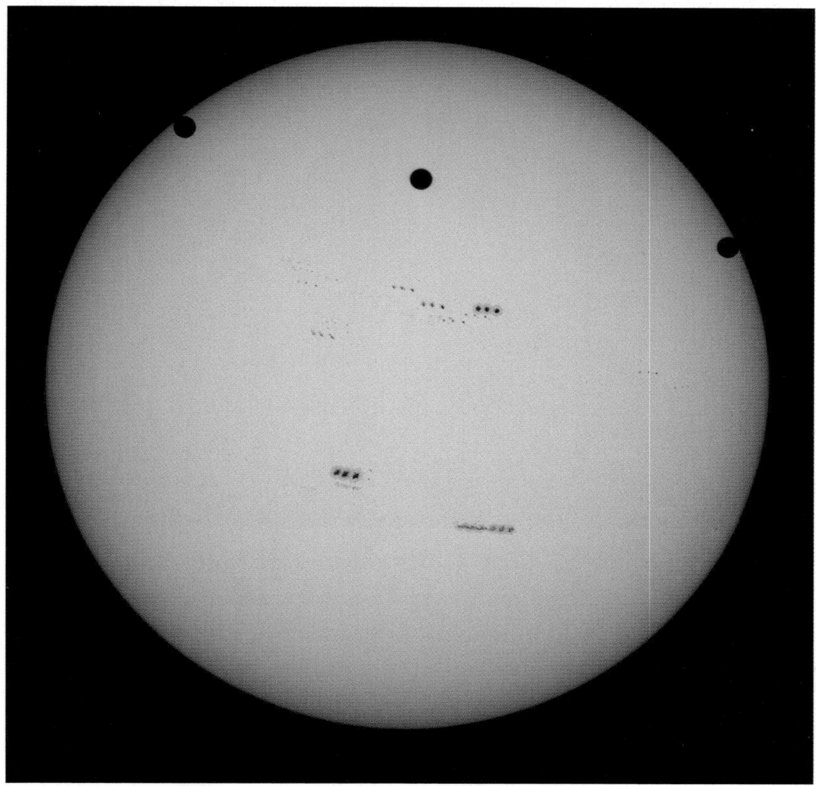

PLATE XV. An overlay of three observations at 3-hour intervals of the 2012 transit of Venus observed from Hawaii. Note that the solar rotation during those intervals means that the sunspots are shown in three different positions.

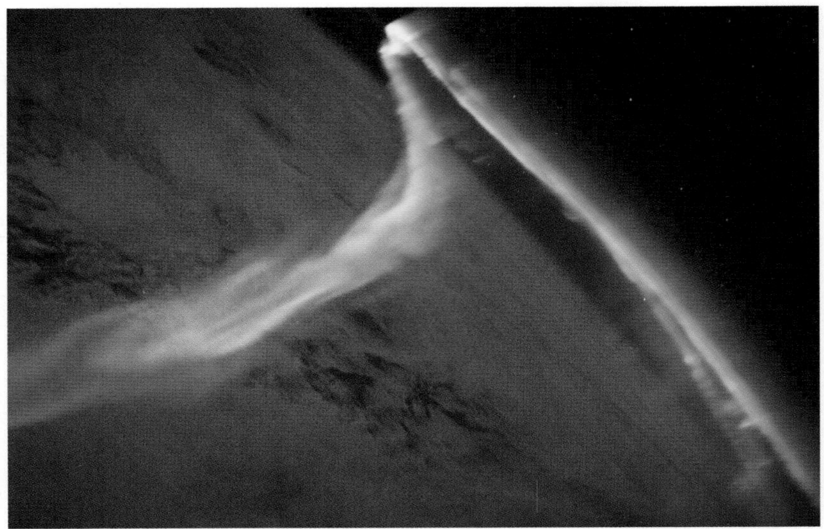

PLATE XVI. On May 29, 2010, looking southward from about 350 kilometers above the southern Indian Ocean, astronauts onboard the International Space Station watched this enormous, green ribbon shimmering below. Known as aurora australis, the auroras on May 29 were likely triggered by the interaction of the magnetosphere with a coronal mass ejection that erupted from the Sun on May 24.

1

The Sun

ORIGINS

About 13.8 billion years ago, for reasons we do not yet understand, the Universe came into existence. Matter as we know it did not exist and even the forces by which bits of matter and radiation interact with each other were different than they are today. Our knowledge of physics is good enough now for us to calculate the conditions prevailing back to an incredible 0.000000000000000000000000000000000000000000001 seconds (10^{-45} s) after it all started. Of course, this does not get us all the way back to zero or before (if the word "before" has a meaning in this context), but we think we can speak with a fair degree of confidence about how things proceeded thereafter.

By 0.000000000001 seconds of age, the four forces of nature that now exist – gravity, strong and weak nuclear, and electromagnetic – were in place, and by the age of several hundred seconds the Universe contained the familiar matter that continues to exist today, the stuff of which ordinary atoms are made. A major turning point occurred at the age of about 380,000 years, when the Universe cooled enough for electrons to combine with the available nucleii, which were mostly protons and helium. At this point, atoms started to form and it suddenly became possible for photons of light to travel long distances without being absorbed. Before this time the Universe was opaque, and our best telescopes will not be able to look back beyond this era.

Some time later, at an age of half a billion years, galaxies started to form. Until then, there were few stars and therefore no sources of light – the universe was in a dark age. Since galaxies consist of large numbers of stars, this implies that many billions of stars were forming, and we must assume that some fraction of them had planets as well. The Universe since then has changed only in some details – galaxies have evolved, the fraction of matter in heavy elements has increased a bit – but has otherwise looked pretty much the same as it does now.

The formation of the Sun is one extremely minute part of this history, the story of one tiny star among the trillions that have come and gone during the past 14 billion years. It is a relatively young star, only 5 billion years old and thus not of the first generation. This means that it, and the planets around it, contain heavier elements formed when earlier stars became novae and supernovae. These heavier elements – oxygen, silicon, iron, carbon, and so on – make possible certain side effects, such as organic life.

The Sun

The Sun is by far the brightest object in our sky, and the difference between its presence or absence overhead is literally like night and day. It is clearly far away, although it took centuries to figure out just how far. How is it, then, that we can know anything about an object that is far away, extremely hot, and astoundingly large?

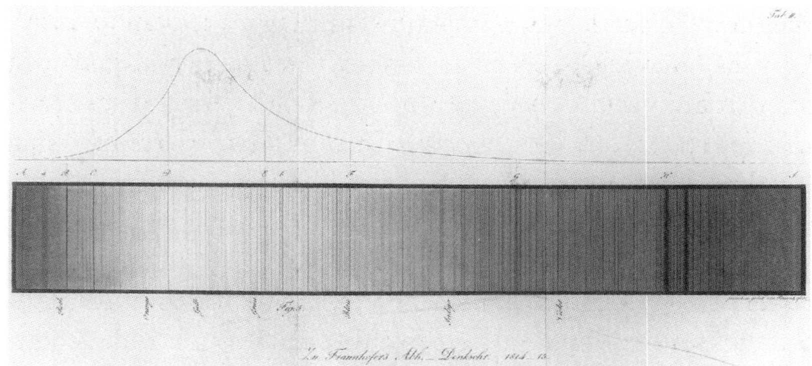

FIGURE 1.1. Fraunhofer's photospheric spectrum published in 1817. This was the first spectrum to show sharp, dark, absorption lines in the solar light. The locations and strengths of these numerous "spectral lines" provide information about the physical conditions on the Sun.

The answer is that the information is in the light. The science of spectroscopy allows us to analyze the solar light in detail (see Fig. 1.1 for an example) and thereby learn about the elements that compose the Sun and their physical states. If we then also use high resolution images of the Sun, we are able to find out what physical processes are occurring to produce the type of light that we see. We can even, with the help of new methods of measurement, now study the interior of the Sun as well.

There is more than light coming from the Sun. Extremely high energy particles from large solar eruptions sometimes reach the surface of the Earth and are detected by terrestrial monitoring equipment. By putting instruments into space, we can extend the range of wavelengths available, enabling us to see solar phenomena not visible from the ground, and we can also intercept and study some of the actual solar material as it flows past the Earth at hundreds of miles per second.

Despite a rapid accumulation of new knowledge in the past 20 years, there is still a great deal we don't know about the Sun. The gaps in our knowledge have broad implications, because solar studies are relevant to almost all of astrophysics: many of the more exotic aspects of astronomy concerning distant stars and galaxies must, of necessity, be based on a foundation of theories and models developed and tested in the solar context. Our ability to explore the unfamiliar territory of intergalactic space reflects how well we understand the more familiar object close to home.

The Sun's physical parameters

This chapter and the next two will give the reader a taste of the scientific process by looking at some of the fundamentals that must be understood before more detailed discussions are possible.

A tabulation of the Sun's basic physical parameters and those of the major solar system objects is given in Appendix III. But these numbers are so far removed from our ordinary experience that they are hard to picture in a meaningful way. To make the data more accessible we will use ratios and analogies. Here are some of the basic facts:

- The ratio of the Sun's diameter to that of the Earth is: 109
- The ratio of the Sun's mass to that of the Earth is: 333,000
- The ratio of average Solar density to that of the Earth is: 1/4
- The ratio of Sun's mass to the sum of all the masses of all the planets is: 744

What do these numbers mean? The Sun is *big* by Earth standards, over a hundred times the diameter, meaning more than a million times the volume. The smallest features that we can see on the Sun with the naked eye or with low-power telescopes, such as sunspots, are typically about as big as the Earth.

The Sun is also very massive, having over three hundred thousand times more total matter than the Earth. Since we said that the Sun is a million times bigger than the Earth, then if it had the same density as the Earth, it would be a million times more massive. But its density is low, only one-fourth that of the Earth, giving it about the same average density as water. The implication of the low density is that the Sun is not made of the same stuff as is the Earth. It is mainly made of hydrogen[1], the lightest element, followed by helium, the second lightest element.

The Earth is made mostly of heavier elements, with very little hydrogen or helium, even though modern cosmology tells us that these two light elements are the most plentiful by far in the entire universe. It would seem that during the formation of the solar system, something caused planets like Earth to end up with more heavy elements, or with less of the lighter elements. Today's explanation is that the smaller planets such as the Earth did not have enough gravitational pull to hold onto very much hydrogen; it escaped back into space and we ended up mainly with the relatively rare heavy elements – oxygen, silicon, magnesium, and iron being the most abundant. The large planets in the solar system, such as Jupiter and Saturn, have stronger gravity and were able to hold onto the light elements, so they have far lower density than the small inner "rocky" planets.

The fourth datum on the list explains why the Sun is the center of our solar system: it has over 700 times as much mass as all of the solar system planets combined, including comets and asteroids. All of these objects form a self-gravitating system: floating freely in space, they are held together by their mutual gravitational pulls and are relatively uninfluenced by other distant masses. In such a system, if one of the masses is much larger than the others, then it will be nearly unmoved by the gravitational pull that the other bodies exert on it. For our solar system, the Sun has about 99.9% of the total mass. This means

[1] That hydrogen is the most abundant element was first realized by Cecilia Payne in her 1925 Radcliffe College dissertation, but it was so contrary to the expectations of the time that, under pressure, she labelled her result "spurious."

that in the gravitational tug-of-war among all of these bodies orbiting around each other, it is a very good approximation – to an accuracy of about 0.1% – to say that the Sun remains stationary at the center of the solar system and all of the planets, asteroids, dwarf planets, Kuiper-belt objects, comets, and so on orbit around it.

The brightness of the Sun

We start by trying to figure out just how bright the Sun really is, how much energy it emits. We can get some idea by making measurements on Earth, measuring the brightness here and also figuring out how far away the Sun is.

The technique is this: assume that the Sun radiates equally in all directions[2], so our local data are representative of what anyone, anywhere, at our distance from the Sun would measure. This distance defines a spherical surface enclosing the Sun and having radius equal to 1 au (redefined by the International Astronomical Union in 2012 as exactly 149,597,870,700 meters), the distance between the Earth and the Sun. If we then take our measured value and multiply it by the surface area of this enclosing sphere, we will capture all of the radiation emitted in all directions and we will have determined the total power emitted by the Sun.

In order to make this calculation, we need to know the radius of the sphere, by finding the distance between the Earth and the Sun. The problem is that we have no obvious way of finding this number, since no direct measurement is possible. What we can do from our terrestrial location is to look around the sky at the objects out there - Sun, Moon, Venus, Jupiter, and so on - and to measure angles between them. We can use these data to figure

[2] A famous joke about using simplified models has a theoretical physicist calculating milk production by starting with the words "Assume a spherical cow...."

out *relative* distances: by measuring angles we can lay out the geometric pattern of objects in the solar system and determine, for instance, that the Sun is 400 times as far away as the Moon. The solution then is to know the distance to either one of them, because then we know the distance to both.

The first person known to have made such a measurement was Aristarchus in the 3rd century B.C., who measured the angle between the Sun and Moon when the Moon was exactly half full. If the Sun were infinitely far away, this angle would be 90°; with the Sun at a finite closer distance, the angle is slightly smaller, as shown in Figure 1.2. Using this method, the number we calculate for the distance of the Sun is extremely sensitive to the measured value of the angle. Aristarchus measured 87 degrees, whereas the value is really more like 89.85 degrees. His calculation said that the Sun is 18 times further away from the Earth than is the Moon, rather than 400 times further as it really

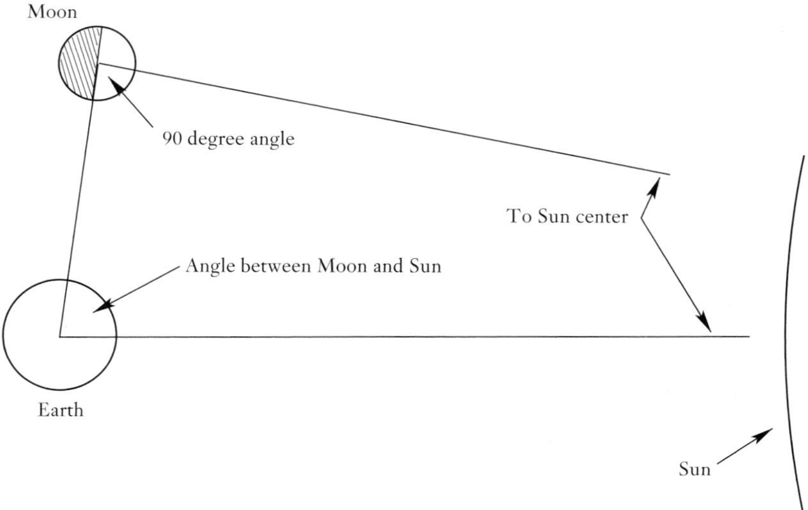

FIGURE 1.2. The geometrical relationships between the Earth, Moon, and Sun used by Aristarchus to determine the distance of the Sun.

is. Still, the method he used was sound, and with more accurate measurements of this sort we can determine the *relative* distances of the Sun, Moon, and planets quite well.

But we still do not know the true size of this pattern of relationships. How do we ever find any *absolute* distances? The answer to this problem turned out to require the invention of extremely accurate clocks.

Transits of Venus

Why clocks? Because the method used was triangulation from widely separated points on the Earth, whose size we know from measuring it directly, and the method requires an absolute determination of when the measurements are made.

In astronomy we speak of *parallax*, rather than triangulation, to denote the well-known phenomenon that two objects line up differently along the line of sight for one observer than for another. For example, during a solar eclipse an observer at one location might see the Sun and Moon line up perfectly, so that the eclipse is total. But an observer some distance off to the side might be able to see past the edge of the Moon for a partial view of the solar disk; for her, the eclipse will not be total.

We have known the relative distances among the planets for quite some time – the values have not changed much since the days of Copernicus. We have also known how fast the planets move around in their orbits, so that the angles between them and how these angles change with time has been known for many years. But in order to progress from a relative diagram to one whose absolute size is determined, we need to know the true length of any one piece of the figure; we will then know all of the lengths. Parallax can be used to make this measurement.

We do have access by direct measurement to one length: the size of the Earth, or more to the point, the distance from one side of the Earth to the other. All we need then is to line up two objects from one side, then from the other, and measure the size of the angle between these two perspectives, and we know the absolute scale of the solar system. But what to use?

In 1716, Edmond Halley – who not only plotted the orbit of the comet that bears his name, but who also was the first accurately to predict the path of a total solar eclipse – pointed out that the passage of the planet Venus across the face of the Sun could be used to provide the needed marker. There had been a pair of transits of Venus in 1631 and 1639; the first went unobserved and the second, though observed, did not lead to useful measurements. The next pair of transits would occur in 1761 and 1769 and Halley, knowing he would not live to see them, urged future astronomers to make the extraordinary efforts needed to obtain the crucial measurements from widely separated parts of the Earth.

Time enters into the measurement because the contact between Venus and the bright disk of the Sun occurs at different times at the two separated sites. If we imagine that the line joining the edge of the Sun to Venus is continued out until it hits the Earth, then this line sweeps across the Earth as Venus moves across the face of the Sun. First it hits one edge of the Earth (observer number 1), then it sweeps across until it hits the other edge of the Earth (observer number 2). From our scale model – whose absolute size we are trying to determine – we know the rate at which this line is sweeping around. A measurement of the time between the transits seen at the two sites therefore tells us how much angle has been swept out in the time it takes to sweep from one side of the Earth to the other. Knowing that angle and the size of the Earth, which is one side of the triangle, allows us to calculate the size of the side of the triangle formed by the distance between the Earth and Venus. The key is for the two observers – who in those days could not communicate with each other – to both have highly accurate clocks that tell the same time if only one contact is measured, or, by Halley's original method, clocks measuring time at the same rate so as to determine the duration of the transit.

The entire business of getting to the appropriate sites, bringing accurate enough clocks, avoiding hostile natives, obtaining the data, and getting home safely was a major international undertaking[3] but it was accomplished and the distance from Earth to Sun was determined though not as accurately as desired.

The nature of the orbits of the Earth and Venus is such that transits of Venus come in pairs with 8 years between them, and then over a hundred years until the next pair. There weren't any transits of Venus since 1874 and 1882 (Fig. 1.3) until the pair of transits on June 8, 2004, and on June 5-6, 2012 (Plate XV). The 2004 transit was visible from all of Europe, with part of it visible from the Eastern United States, and entirely visible in western Siberia and in progress at sunrise in eastern Russia. A whole transit of Venus observed from Earth, from immersion of Venus onto one side of the Sun until emersion on the other side, lasts over 6 hours. The following pair of transits won't occur until December 11, 2117, and December 8, 2125. None of these transits are dramatic to see the way eclipses are, but intellectually they are superb.

Transits of Mercury also occur. They are more frequent than those of Venus, coming about 13 times per century, but are more difficult to observe because Mercury is so small. The TRACE satellite observed such a transit toward the end of 1999 (Fig. 1.4)

[3] The story of the transit expeditions in the 18th century is told in the delightful and often hair-raising book *The Transits of Venus*, by Harry Woolf, Princeton University Press, 1959. See also the series of articles by Don Fernie in *The American Scientist*, and books by Sheehan and Westfall, Maor, and Lomb listed in the bibliography at the back of this book. The two authors of this book and Glenn Schneider solved the problem of the black-drop effect that diminished the accuracy of transit measurements, reported in the journal *Icarus* in 2004.

FIGURE 1.3. The transit of Venus of 1882, photographed at Vassar College by Maria Mitchell and her students.

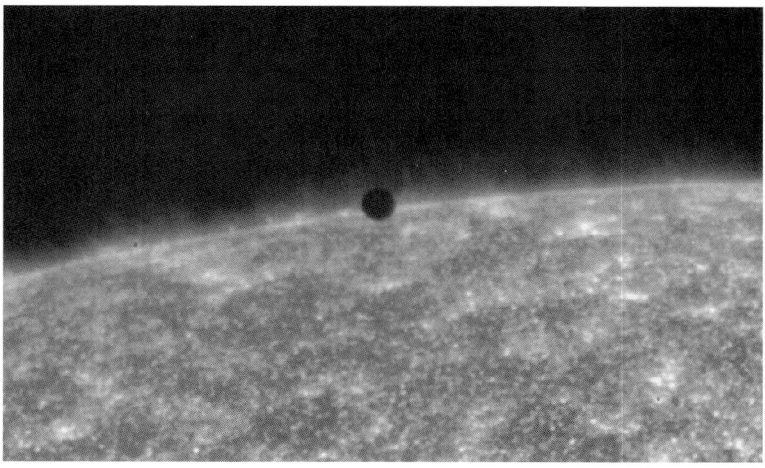

FIGURE 1.4. The transit of Mercury on November 15, 1999, observed from Earth orbit by the TRACE satellite.

and also in 2003 and 2006. The next Mercury transit as viewed from Earth is May 9, 2016.

Luminosity: the brightness of the Sun

Determining the Sun's distance from the Earth with any reasonable accuracy was a major achievement that required much ingenuity and many centuries of work. Once the distance to the Sun is known, we can determine the Sun's total brightness, as described above: we put our detector in the sunlight, measure the amount of energy falling on the detector and then calculate what would have been measured by a detector completely enclosing the Sun.

The detector must be one that takes into account all wavelengths, including the invisible infrared and ultraviolet. Such a device is called a bolometer, and the brightness measured this way is called bolometric luminosity. It is best if we place the bolometer above the Earth's atmosphere, so that the absorption of sunlight by air is avoided. Fortunately, this can be done these days, and the solar constant (which, as we'll see, turns out to be not really a constant) can be measured to an accuracy of a small fraction of a per cent.

The number turns out to be staggeringly large by human standards: the Sun emits 380,000,000,000,000,000,000 megawatts of power. (Note that power has dimensions of energy per unit of time; one watt is one joule per second.) In one thousandth of a second, the Sun emits enough energy to provide all of the world's current energy needs for 5000 years. Unfortunately, we have no way at present to harness this energy, except for a minuscule fraction of it by using devices such as photovoltaic cells, solar collectors, wind farms, and – indirectly – the burning of fossil fuels.

Energy from the solar interior radiates out into space from each part of the solar surface. The total power emitted by the Sun can be calculated by multiplying the intensity of the emission from each piece of its surface with the total surface area. We can also do the inverse of this calculation, in which we know how much power the Sun emits (for example, by the type of measurement described above), and we know how much surface area it has, to then calculate how much power is being emitted by each piece of its surface. This in turn allows us to estimate the temperature, since the radiating power of a surface depends very strongly on its temperature. In order for the Sun, with a surface area of six billion billion square meters, to radiate away 380,000,000,000,000,000,000 megawatts, it must have a temperature of about 6,000 K, or 10,000°F.

HOW WE STUDY THE SUN

Scientific instruments extend our observing range beyond the capabilities of our senses: we make distant objects clearer with a telescope, or small objects visible with a microscope; we build instruments that can see infrared, radio, ultraviolet, and X-ray wavelengths; we build particle detectors to measure cosmic rays or reaction products from the collisions of high energy particles; and so on. The possibilities are endless, and we will give just a small sample of the types of instruments that are used today to study the Sun.

Telescopes

Telescopes that image the Sun have special problems not encountered in night-time astronomy, mainly caused by the extreme brightness of the Sun. First, there is the well-known problem that one should never look at the Sun directly through the eyepiece of a telescope, since this would cause major damage to the eye. So a solar telescope either must be equipped with some type of filter that reduces the brightness of the image, or the image must be projected onto a screen for indirect viewing.

FIGURE 1.5. The New Jersey Institute of Technology's Big Bear Solar Observatory, with its second-generation dome hiding its 1.6-m New Solar Telescope, on Big Bear Lake, California. The observatory was built in the 1970s by Harold Zirin of Caltech, and opened in its new configuration in 2009.

Electronic recording via a CCD camera with suitable exposure times is also a possibility.

Selection of the observing site is of major importance for solar, and indeed for all, telescopes because the amount of turbulence in the atmosphere has a major impact on the quality of the images obtained. Some of the best sites are surrounded by water, such as the solar telescopes in Hawaii and the Canary Islands, or the Big Bear telescope, situated on an artificial island in Big Bear Lake, California, now connected to the shore by a causeway (Fig. 1.5).

Once these problems are overcome, another set of problems is encountered. The light from the Sun heats the mirrors, causing them to deform and ruining the image quality. For instance, one of the standard telescope designs is a Cassegrain, in which a large mirror (the primary) focuses light onto a smaller mirror (the secondary) in order to produce a highly magnified image in a fairly compact system. A typical magnification factor for the secondary mirror in a Cassegrain is 6x, which means that the intensity of light falling on the secondary is roughly 36 times as large as that falling on the primary mirror (36 = 6×6). For starlight this is not a problem, but for sunlight the 36 solar constants of intensity hitting the secondary mirror amount to nearly 50,000 watts per square meter. Even if the mirror has a highly reflective coating, absorption of only a few per cent of this intense light will cause severe heating and distortion of the optical surface.

It also may happen that the solar image being focussed by the mirrors heats the air inside the telescope tube, since those same thousands of watts are passing through the air. Air is not 100% transparent, and absorption of even a small fraction of the solar energy will heat the air, which then starts to expand and convect away from the heating site, resulting in local atmospheric turbulence inside the telescope that again ruins the image. To solve this problem, which is peculiar to solar telescopes, several sites have built a "Vacuum Tower Solar Telescope," which means that the entire telescope interior – roughly the size of an elevator shaft – has been made into a vacuum chamber. The large solar tower in Sunspot, New Mexico, shown in Figure 1.6, is such an instrument. Note that the entire telescope is suspended on a trough of liquid mercury (labelled Float in the diagram) in order to reduce vibrations and so the central structure can turn to keep up with the image rotation provided by the use of a heliostat, and that the observer stands on a platform attached at ground level, about two-thirds of the way from the base up the tower (labelled 40 Foot Dia. Table), in order to have access to the focused image. This facility is part of the National Solar Observatory, which also operates the McMath-Pierce Solar Telescope – the largest solar telescope in the world, until the size is superseded

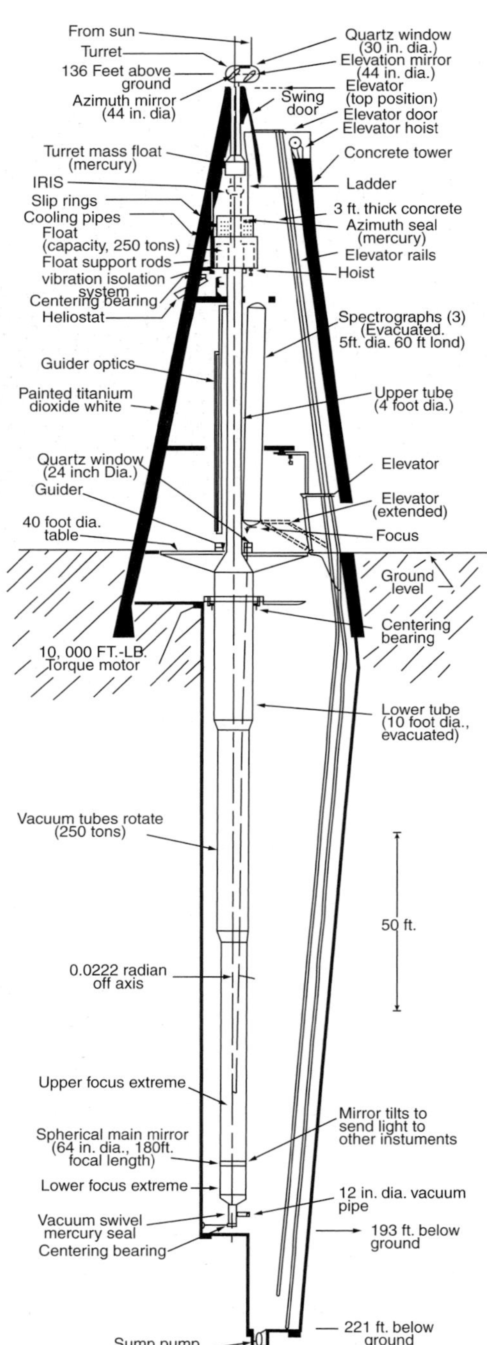

FIGURE 1.6. Diagram of the Richard B. Dunn Solar Telescope at the National Solar Observatory's Sacramento Peak Observatory in Sunspot, New Mexico. Note that light travels down 329 feet (100.3 m) in vacuum to the primary mirror 188 feet (57.3 m) underground. The entire 200-ton (≈180,000 kg) central structure and observing platform is suspended near the top in a 10-ton (9000 kg) tank of mercury.

FIGURE 1.7. The world's largest solar telescope (along with Big Bear's newer New Solar Telescope and until the Advanced Technology Solar Telescope comes online), the McMath-Pierce of the National Solar Observatory at Kitt Peak, Arizona. The telescope is the triangular structure in the foreground; a vertical tower holds up the inclined path (oriented toward the north celestial pole) into which a heliostat at the top directs the incoming sunlight down below ground to the 1.6-m primary mirror.

by the Advanced Technology Solar Telescope (ATST) – at Kitt Peak just outside of Tucson, Arizona (Fig. 1.7). This Daniel K. Inouye Solar Telescope represents a collaboration of 22 institutions, reflecting a broad segment of the solar physics community. The construction phase of the project, to build this next generation ground-based solar telescope at the Haleakala High Altitude Observatory site, is currently under way.

In addition to these telescopes operating in visible and infrared light, there is also much to be gained by studying the Sun at very long electromagnetic wavelengths, in the radio portion of the spectrum. The first detections of the Sun at radio wavelengths were outbursts generically called "noise storms" (described below) but the more normal non-flaring Sun can now be imaged as well.

One of the problems of building a radio telescope is that, in order to have sharp images, a mirror must be a certain size that depends on the wavelength of the light being imaged. This is not a problem in the visible, where the wavelength is a tiny fraction of a hair's width, and mirrors only a few inches in diameter can begin to give high quality images. But radio waves are thousands of times longer than visible-light waves – several inches to several feet between peaks. The corresponding telescope diameter needed for imaging the Sun would be hundreds to thousands of feet, and it would need to be pointable, since the Sun moves across the sky.

Such a device is not practical to build, so that clever alternative methods must be used, the main one being arrays of smaller antennas arranged so that they can be treated as if they were a single large antenna. For instance, one of the first such "radioheliographs" was an east-west linear array of 32 aerials stretching in a line nearly a mile long, constructed at Nançay in France in the 1950s. Such an array is long in one dimension and short in the other, so the resolution of the array is good in one dimension only. The Culgoora radioheliograph in Australia (no longer in operation) was built as an array laid out on a 2-mile diameter circle, thereby providing two-dimensional, rather than just one-dimensional, images.

The individual elements of such arrays must be properly spaced and the signals from the elements must be carefully combined, but the technique works and is being used for radio imaging of the Sun. Another 2-D arrangement is a crossed pair of linear arrays known as the Mills Cross; this arrangement was used at the Clark Lake, California, radioheliograph. The Jansky Very Large Array (JVLA), in Socorro, New Mexico, was not built specifically as a solar instrument, but it is used during a portion of its observing time to acquire high quality

FIGURE 1.8. An aerial view of the central part of the Jansky Very Large Array, looking toward the South.

solar radio imaging observations (Fig. 1.8), and its upgrade extended its solar usability. Another facility dedicated to solar radio observations is the Nobeyama Solar Radio Observatory (Fig. 1.9). A Chinese radioheliograph opened in 2012 in Inner Mongolia.

Spectroscopy

Before we get to a discussion of spectroscopy we need to make a detour into the nature of light and color. Historically, humans have had great difficulty in understanding what light is. Is it something emitted by burning objects, or does the burning activate the surrounding space in some way? Does the light fill all of space instantaneously, or is time required? Do our eyes emit some kind of ray that reaches out and senses an object, or does something from an object enter our eye? If the latter, how does

FIGURE 1.9. The Nobeyama Radio Heliograph.

the shape, size and distance of an object register accurately on our senses? What is color? Is it something separate from light?

The properties of light that we find so puzzling are almost all due to the following result of quantum mechanics: light consists of particle-like bundles called photons, which can be created and destroyed, and which have zero rest mass. Light does not exist "in" a fire, but is produced by the hot atoms of the fire. Light does not flow into our eyes and stay there, like water poured into a glass, but disappears – ceases to exist – when it hits the photoreceptors of the retina. Light therefore seems more ephemeral than, say, molecules of water, which can evaporate from the surface of a pond, or fall as rain and collect, but the total number of which remains constant.

Light also travels very quickly by human standards. Because of its zero rest mass, a photon of light travels at the maximum possible speed, which (no great surprise) is known as the speed of light. This speed, roughly 186,000 miles per second, is so great that it might as well be infinite, as many ancient people thought, since high-precision equipment is required to make the measurement. (Galileo tried, by sending a friend to a distant mountain with a lantern and having him uncover his lantern when he saw the light from a lantern uncovered by Galileo. Good experimenter that he was, he reported that the speed was too great to determine by this method, which is exactly right.)

In addition to being so ephemeral, light also has the puzzling property of being associated with color. But we do not see color flashing past us in the air, so it must be something apart from light. Is it a property of objects, that is revealed by light? Is light one thing, and color another? (Technically, that is the case for standard TV broadcasts, which have separate "luminance," i.e., brightness and "chrominance," i.e., color, pictures.)

That white light can be spread out into colors has long been known. Newton carried out a long series of experiments in the 1660s using sunlight and prisms and showed that incoming white light will form a spectrum when passed through a good quality prism. From the observation that the different colors are bent by different amounts in the prism (Fig. 1.10), and

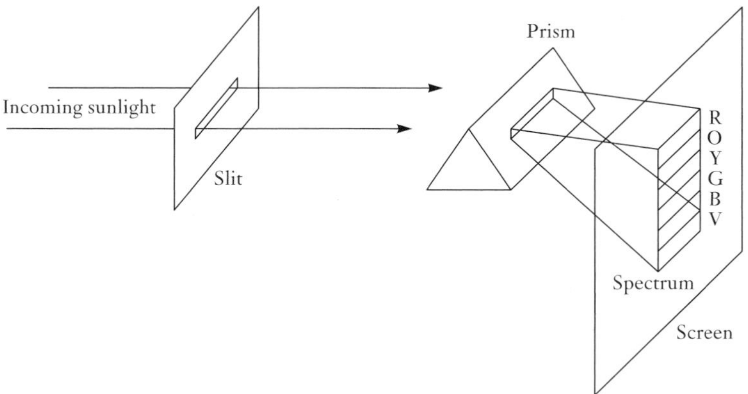

FIGURE 1.10. Using a prism to split light into its component colors

that those colors are then bent the same amount when passed through a second prism, he concluded that white light is made up of "difform Rays, some of which are more refrangible than others."

It was William Wollaston in 1802 who commented on the "gaps" between what he saw as four colors, with some lesser gaps also indicated. Instrument-making techniques were then rapidly evolving and one of the opticians who improved the techniques was Josef Fraunhofer in Germany. He used a spectrograph with a slit, which prevented the overlapping of colors from different parts of an extended source. In 1811 he produced an epochal chart of the solar spectrum, which is a continuous band of color and shows hundreds of dark lines across it, of which he drew only the strongest, as shown in Figure 1.1. We now know that this continuous spectrum with absorption lines is from the solar photosphere. Explaining the origin of these lines led physicists, and will now lead us, to the quantum theory.

Every element has its own unique arrangement of electrons. These negatively charged electrons (thus called from the arbitrary sign convention established by Benjamin Franklin) are drawn to the positively charged nucleus according to the number of charges in that nucleus, and normally just enough electrons are attracted to make the atom as a whole neutral. For instance, carbon, with six positively charged protons in its nucleus attracts six electrons which fall into place in a very specific arrangement around the nucleus. There is no consideration of one type of electron or another, because they are all the same.

This is a crucial point, because it means that on the atomic level, every atom of neutral carbon is *exactly* the same as any other one (except for being at a different location in space). This sort of thing does not happen in the "macro" world, the world of our ordinary experience. In this macro world, things can be "pretty much" the same, but are extremely unlikely to be truly identical. But identity at the quantum level has a bizarre (to us) consequence: when two identical bits of matter interact, they can change places, and continue on as if one were the other, and vice versa. This would be truly astounding at macroscopic scales and shows that the word "identical" has a very definite technical meaning in physics.

One consequence of quantum physics is that each type of atom has a very well-defined and unique way of absorbing and emitting light, which is directly connected to its specific arrangement of electrons. The electrons around each atom can exist only in certain discrete configurations[4], which means that they can only absorb and emit energy in certain prescribed amounts. This form of interaction between each type of atom and the photons of light leads to a characteristic pattern of light absorbed by the element or emitted by it.

The pattern of light emitted or absorbed by an element provides a unique fingerprint for the presence of that element. A familiar example is the element sodium, which when heated emits light very strongly at a very narrow and specific range of wavelengths in the yellow part of the spectrum. The presence of small amounts of sodium ("salts") gives many flames a characteristic yellow color.

The spectrograph

As shown in Figure 1.1, the light coming from the Sun can be spread out in wavelength to reveal an enormous amount of detail, in the form of sharp variation in the strength of the light at different wavelengths. One needs to only glance at such a display – also known as a "spectrum" – to guess that there is a great deal of information contained in these variations, if

[4] Think of climbing a ladder: you can only stand on discrete rungs, but there is nowhere to stand in between. Now imagine that there is no in-between: you can only be on one rung or another, but nowhere else.

one only knew how to decode it. Extracting information about the physical state of the object emitting the radiation from a detailed study of the spectrum became one of the main activities of astrophysicists in the 19th and early 20th centuries.

An instrument used to make and record a spectrum is called a spectrograph. When an eye or a piece of film receives light, it records a range of wavelengths all at once. In a black-and-white photograph the wavelengths from red through blue are all added together, while in a color photo some separation is made into three broad bands, such as red-yellow-blue. But in none of these cases is the discrimination fine enough to discern the very narrow (in wavelength) features, known as spectral lines, that will provide the solar physicist with enough information to, for instance, determine the chemical composition of the gas. A special instrument that disperses the light, strongly spreading out the wavelengths, is needed.

To make a spectrograph, what is required is a device that bends different wavelengths of light by different amounts. With such a device, the wavelengths can be spread out spatially so that each wavelength appears in a different location on the detector. Two main methods are used, the prism and the diffraction grating. The prism works as described above in Newton's experiments: different wavelengths are bent (refracted) more or less strongly according to their wavelength, so that a ray of white light entering the prism is split into its component colors when exiting the prism, as illustrated in Figure 1.10.

The other main method used for producing spectra is the diffraction grating. Here, a series of very closely spaced lines is cut into a surface, and the light is reflected from the grooves. When the reflected waves are in phase with each other, the reflection is enhanced; otherwise they cancel each other out. The enhancement or cancelling of each wavelength occurs at a particular angle, so that each wavelength comes out of the grating at a characteristic angle. The result is thus similar – you see a spectrum of colors – to the effect of a prism, although it turns out that gratings can be made to produce a much stronger spreading of the light than prisms can.

In all likelihood, you have such a diffraction grating at home – a compact disk. If you hold its surface to the light you will see a spectrum of colors from the closely-spaced rows of reflecting surfaces used to record data on the disk.

The coronagraph

On rare occasions – roughly once every four centuries at a given spot on the Earth's surface – the Moon passes in front of the Sun so as to block out (eclipse) all of the Sun's visible light. The Sun is so bright that a partial eclipse, in which the Moon does not completely cover the entire disk of the Sun, might not even be noticed without advance warning. Even when 90% of the Sun is covered, the amount of darkening is roughly the same as moderate cloud cover.

But when that last little bit of the Sun is covered, the effect is dramatic. The world suddenly grows quite dark and bright stars become visible in the sky. A kind of glowing aura surrounds the Sun's blackened disk (Fig. 1.11). It was many centuries before it became clear that this aura actually belonged to the Sun, rather than a result of scattering in the Earth's atmosphere, or even an effect of an atmosphere on the Moon illuminated by the bright Sun behind it.

The corona is extremely faint. It is only about one-millionth as bright as the disk of the Sun, and cannot normally be seen. During an eclipse, the bright disk is covered up, and the amount of light scattered by the Earth's atmosphere is reduced enough to allow the bright inner portion of the corona to be seen. How, then, could astrophysicists see the corona at will, without waiting for an eclipse and travelling to some distant location?

FIGURE 1.11. The November 14, 2012, total eclipse, a composite of many images taken from the observing site in Australia.

The enormously difficult problem of viewing the faint corona near the bright solar disk outside of eclipse was solved in the 1930s by Bernard Lyot (pronounced lee-OH). First, a site with exceptionally clear sky must be found, because humidity and air pollution or dust scatter light, causing the brightness of our atmosphere around the Sun to swamp the coronal brightness. Then a telescope is fit with a device that covers the bright solar disk; this is often done with a small circular plate, called an occulting disk which hides the bright part from view, allowing the faint outer part, the corona, to be seen (Fig. 1.12). But the sky is too bright, even from high mountains, to see the corona well. At the 3,440-meter (11,290-foot) level of the Mauna Loa Solar Observatory in Hawaii, the High Altitude Observatory runs a coronagraph that uses polarization to distinguish the corona better. Since the corona is highly polarized and the sky is not, their analysis displays the corona farther out from the sun

(a)

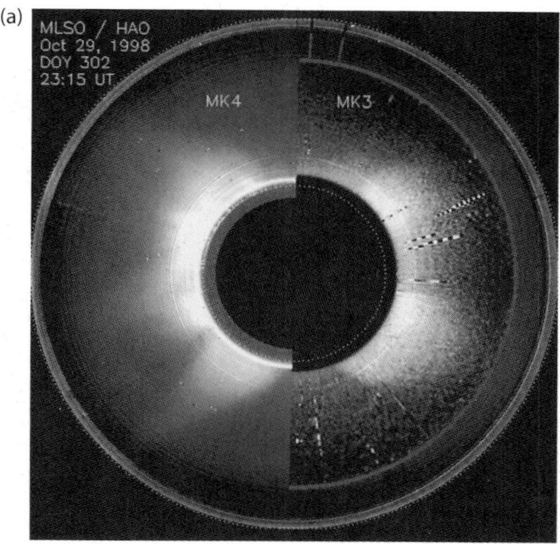

(b)

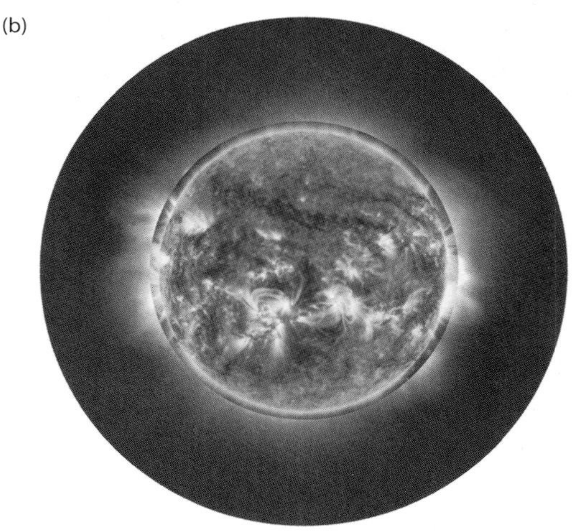

FIGURE 1.12. a) A comparison of the image quality between the older Mk3 coronagraph and its successor, Mk4. That coronagraph was superseded in 2014 by an improved white-light coronagraph called K-cor. b) K-cor, with an engineering-test image from 2013 shown here surrounding an ultraviolet coronal image from the Solar Dynamics Observatory spacecraft. The image shows K-cor's improved imaging of the low corona. It should improve Mk4's speed and signal-to-noise ratio by a factor of about 10 and improve the accuracy of its calibration and its reliability. It will image every 15 seconds from 1.05 to 3.0 solar radii, which should be especially favorable for imaging coronal mass ejections. K-cor will be part of COSMO, the Coronal Solar Magnetism Observatory, a proposed facility with 3 instruments. See http://mlso.hao.ucar.edu/.

than is visible with ordinary coronagraphs. The K-cor coronameter at Mauna Loa has now been put into service, providing greatly improved coronal imaging out to several solar radii from the limb.

But merely covering over the bright disk is not good enough, because the full sunlight still falls on parts of the instrument just outside of the occulting disk's shadow. The rest of the telescope must therefore be built with extreme care given to the light scattered by the lenses or mirrors, by reflections and scattering between one lens and another, and by an inevitable optical effect called diffraction. This last effect is removed by careful positioning of internal baffles (small circular obstructions), known as Lyot stops. A good coronagraph these days will have a reduction of a factor of one trillion in the level of internal scattered light compared to the incoming light.

SUGGESTIONS FOR FURTHER READING

Pasachoff, Jay M. and Alex Filippenko, *The Cosmos: Astronomy in the New Millennium*, 4th edition (Cambridge University Press, New York, 2014).

Eddy, John A., *The Sun, The Earth, and Near-Earth Space: A Guide to the Sun-Earth System*, NASA, Washington, DC, 2009, bookstore.gpo.gov.

Lang, Kenneth R., *The Sun from Space* (Springer, Berlin, 2009).

Zirker, Jack B., *The Magnetic Universe: The Elusive Traces of an Invisible Force* (Johns Hopkins University Press, 2009).

Alexander, David, *The Sun*, Greenwood Guides to the Universe (ABC-CLIO, Santa Barbara, CA, 2009).

Lang, Kenneth R., *The Cambridge Encyclopedia of the Sun*, Cambridge University Press (Cambridge, UK, 2001).

2

The Once and Future Sun

It would be overly optimistic of us to think that we have any accurate understanding of the conditions that make life possible. We do know that life exists on Earth, that it does not seem to have ever existed on the Moon, and that it may or may not have existed on Mars and Venus. If there is life elsewhere, such as on a moon of Jupiter, we have not yet found it. This means that we have exactly one data point, and it is dangerous to generalize from a single example. After all, in recent years we have seen many types of solar systems containing the thousands of exoplanets that have been discovered, and realized that most of them are very different in format from our own solar system. Generalizing from our own solar system turned out to be very wrong.

Still, as far as we can tell, light and heat have been of central importance. The Earth would not have life on it if the Sun had not brought the surface temperature of our planet above the freezing point of water, but well below the boiling point. We are not in a good position to argue that this range of temperatures is absolutely essential for life, but it is generally necessary for the types of life that we see here. Because the presence of the Sun as a source of heat and energy is vital for life on Earth, we may be tempted to ask some pointed questions: why is this object emitting so much light and heat? how long has it been doing this? how long will it keep doing this?

We currently have fairly reliable answers to these questions. We think that the Sun started as a collapsing cloud of gas and dust about five billion years ago, and that it will end up becoming a red giant, expanding in size to engulf the Earth, about five billion years from now. The story of how we came to our present understanding covers a wide range of fields, from geology to atomic physics, and it provides a nice illustration of the meandering and often unpredictable character of scientific progress.

THE EARLY SUN

It has been commented that, in the early stages of development in a field of science – before things are understood – one begins by classifying. Astronomy is no exception, and one way of classifying stars that has turned out to be highly productive is the color–magnitude diagram. Careful examination of nearby stars (where "nearby" is a relative term – the closest is 26 trillion miles away) shows that they are not all the same brightness and color, but rather come in a wide range of sizes. What is most striking, and pleasing, to the orderly mind, is that there is a nice correlation between the brightness of a star – its luminosity – and its color or temperature.

Figure 2.1 is a diagram which shows the observed correlation between brightness and temperature for a sample of more than 16,000 stars. Each dot in this figure represents a star, located horizontally on the figure from its color and vertically from its brightness. When enough stars are plotted we see that there is a large group of stars that fall in a band cutting across the

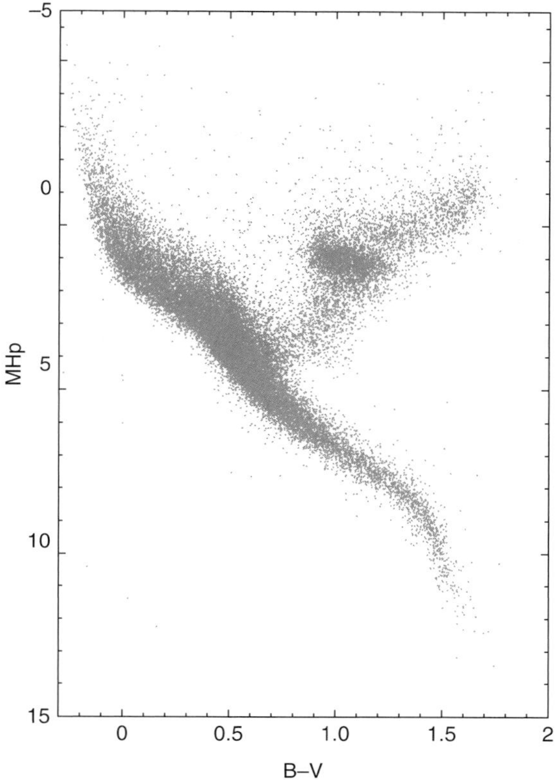

FIGURE 2.1. The Hertzsprung-Russell diagram for 29,606 stars observed with the European Space Agency's Hipparcos mission, based on a 2007 restudy of the data that had been originally published 10 years earlier with improved understanding of the satellite dynamics analyzed with more modern computer facilities. Hipparcos measured the parallax of stars, their slight shifts in the sky when observed from different points in the satellite's orbit, as the most direct possible measurement of their distances from Earth. The distances of these stars, allowing the conversion of apparent (observed) magnitude to the absolute-magnitude scale are known through this study to better than ±10%. The vertical axis is the magnitude (M) measured by Hipparcos (Hp). The horizontal axis is the standard display of blue magnitude (B) minus visual (V, yellow) magnitude, which corresponds to the hottest stars on the left and cooler ones on the right. We now await the analysis of the next generation of observations from ESA's Gaia mission.

diagram, from upper left to lower right. The brightest stars are the hottest and the faintest stars are the most red, or coolest. This swath is known as the "main sequence." (There is another branch containing fewer stars that heads up and to the right, which we will discuss later.)

Such a correlation demands an explanation. We might start by assuming (i.e., guessing) that this narrow band in the diagram means that there is one parameter that is varying among this assortment of stars, but that they are otherwise the same. This would mean that:

1. All of these stars are being powered by the same process, since the pattern forms a continuous distribution across the diagram. Several different processes are very unlikely to connect together in such a smooth manner.
2. The bigger the star, the more strongly the process operates. This is because the brightness of the stars in the diagram correlates with the temperature, with hotter stars being brighter.
3. These stars are all made up of roughly the same stuff, since the band is fairly narrow. Whatever the energy release process, it seems unlikely that different fuels will all produce exactly the same correlation between size and brightness that is observed, so the band would not be so narrow.

Although it is conceivable that many separate stars will all fall into line as observed by accident, or with each one having some different power source inside it, the most economical explanation and the most reasonable is that there is one process operating inside of stars and that it is essentially the same in all of them. The essence of science, though, is not merely qualitative suggestions, but the calculation of quantities. We would therefore need to develop a theory that connects, for instance, the amount of energy generated with the mass of the star; this is our second principle above. Such a theory, in which stars are powered by nuclear fusion, has been developed quite successfully, and we will discuss it in the next section.

The third principle takes into account that the process going on inside of stars to generate energy depends on which atoms exactly are fusing together. With the growing realization early in this century that stars are made up mostly of hydrogen, it was natural to think of hydrogen nuclei – protons – as the main ingredient in this cosmic cooking.

The observations therefore imply that stars begin life by settling down to a long, slow phase of life on the main sequence: if they spent a significant amount of time doing something different, then we would see a large number of them located elsewhere on the diagram. But what happens before and after this main sequence phase? Logically enough, the earlier evolution is called the "pre-main sequence" phase, and it is the time when the cloud of gas and dust is contracting and heating until the center becomes hot enough to ignite nuclear fusion. The later stage, for a star like the Sun, is called the "red giant" phase, which forms the upper-right branch in the diagram. As the name implies, this means that the Sun will eventually go through a change that increases its size enormously, and it will swell to about the size of Mars' orbit. We have roughly 5 billion years to prepare for this disaster.

The age problem

The 19th century saw the rise of geology as a science and continual improvement in the methods for establishing the age of the Earth. At the beginning of that century there was not much reason to doubt the biblically-established age of about 6,000 years. By the middle of the century it was clear that the Earth was at least several million years old, and by the end of the century an age of many millions of years seemed likely. By the start of the 20th century, the natural radioactive decay of elements had been discovered and this eventually allowed an accurate dating of nearly five billion years for the age of the Earth (see Chapter 7).

Because the Sun was presumably shining during all of these billions of years, it became problematic to determine the source that could produce such an enormous amount of energy over such a long span of time. It was known that ordinary chemical processes could not keep the Sun burning more than a few thousand years – an easy calculation that merely takes the amount of mass in the Sun and assumes that it is, for example, a lump of coal that burns – so more exotic processes had to be explored.

In the 1840s, a German physician named Julius Mayer sent a paper to the Paris Academy proposing the energy released by meteors falling into the Sun as the source. There is not a huge amount of meteoritic material falling out of the sky, but the amount of energy released by each bit of matter in falling into the Sun's very strong gravitational field is large, so this mechanism was at least conceivable. A Scottish engineer in Bombay, John Waterston, also submitted a paper to the Royal Society of London proposing that a contraction of the Sun under its own gravitational pull was the energy source. Again, the very large gravitational pull of an object as massive as the Sun makes this a viable mechanism as well. Both papers were rejected.

Waterston continued to work out his theory and in 1853 he gave a talk to the British Association for the Advancement of Science, in which he presented a general discussion of the formation of the solar system out of a rotating disk of gas (an idea that had been discussed by the philosopher Immanuel Kant and elaborated by the mathematician Pierre Simon de Laplace in the previous century), and then went on to talk about the energy source of the Sun in the context of gravitational forces. Waterston included both meteor infall and gravitational contraction in the discussion, and the talk influenced two major figures in the physics world, William Thomson of Glasgow (later Lord Kelvin) and Hermann Helmholtz of Königsberg.

Thomson favored the meteor infall theory and went on to calculate, based on an estimate of the number of meteors still floating around the solar system, that the Sun could last for only another 300,000 years. Helmholtz, working within the view that the Sun had contracted out of a cloud of gas and dust, calculated that the Sun could still be contracting slowly and that this could supply the energy needed to power it for millions of years.

Both of these mechanisms turned out to be inadequate once it became clear that the Earth was billions of years old. As it happened, Helmholtz had also mentioned that the lifetime of the Sun could be longer if an altogether new source of energy were discovered, and he lived long enough to hear Ernest Rutherford proclaim, in the early years of the 20th century, that the discovery of radioactivity provided just such an answer. It was not until the 1920s and 30s that this new source of energy became known and was understood well enough to apply it to the problem of powering the Sun.

HOW THE SUN FORMED

Occasionally, in the vast spaces between the stars of our galaxy, some regions of gas become slightly denser than the surroundings, perhaps from a shock wave resulting from a nearby supernova or perhaps randomly. The interstellar matter includes not only the basic hydrogen of the universe but also some of the modestly heavier elements (like carbon) gently ejected by stars or even the heaviest elements spewed out by exploding stars. (The first three minutes or so after the Big Bang witnessed the formation of hydrogen with all its deuterium, some of the helium, and little of any other type of atoms; additional helium is formed in ordinary stars.) Some regions are Giant Molecular Clouds, including not only atomic forms but also a variety of molecules. By interstellar standards the density of matter in these interstellar clouds is enormous – over a millionth of an ounce per cubic mile – and if the cloud is large enough it begins to collapse in on itself because of its own gravitational pull. Small clouds, less than about a trillion miles across, do not have enough gravity of their own to collapse against the countervailing outward pull of the other stars and galactic material swirling past. But the larger clouds will continue contracting for a million years or so and the material in the middle of the cloud will get hotter as the gas and dust are compressed by gravity.

If the cloud is too big, the temperature and pressure in the middle become so large that the cloud explodes, tearing itself apart. If the cloud is too small, it merely collapses and radiates away the heat of compression over a few million years. But clouds that are in the right range – from a few times as much mass as Jupiter to about 100 times as large as our Sun – do something quite spectacular: they begin converting light elements to heavier elements, by nuclear burning, releasing enormous amounts of energy. The amount of energy generated is so large that these objects – stars, as we call them – can be seen a million billion miles away. And if a rock the size of the Earth is placed about 100 million miles away, the heat from the star is sufficient to warm the surface enough to sustain life.

Nuclear burning

The basic energy-release process in stars is often called "nuclear burning," in analogy with the chemical process of oxidation that we call "burning." In ordinary burning, atoms of oxygen come together with atoms of whatever is burning – say carbon – to form a molecule like carbon dioxide. When the atoms combine, the electrons that were orbiting the separate atoms become shared between the two types of atoms, and the result can turn out to be a configuration that is more stable than the original separated state. "Stable" means that some energy has been released when the two came together and this release of energy produces the heating that we associate with the word burning.

Such rearrangements of electrons among atoms are chemical reactions, and they proceed at energies comparable to those needed to move electrons from one atom to another, namely a few volts per atom. (This is why batteries, which work via chemical reactions, produce a few volts.) Nuclear "chemistry," in comparison, involves reactions that change the state of the nucleus, and this typically requires *millions* of volts. Figure 2.2 shows such a reaction, the combination of two protons to form deuterium. The reason that nuclear processes involve such huge energies is that the force holding the nucleus together is much stronger than the electromagnetic force that holds the electrons in place. The amount of energy involved when a force acts depends on the strength of the force, and the millions of volts

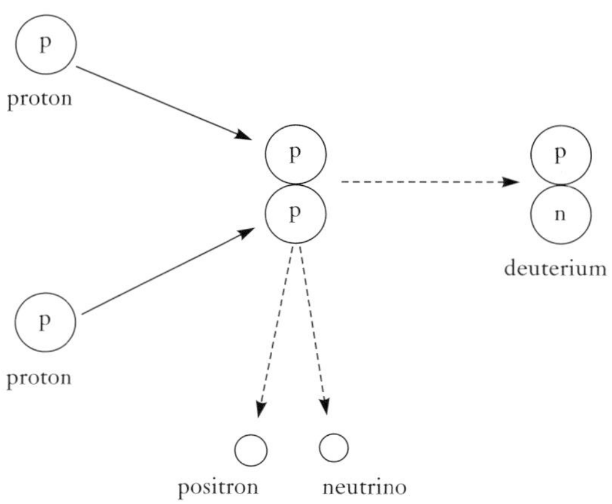

FIGURE 2.2. The first step in the proton-proton chain that powers the Sun. Two protons come together to form deuterium, via fusion, followed by beta decay in which a positron and a neutrino are emitted by a proton, changing it into a neutron.

needed for changes in the nucleus is a reflection of the strength of the nuclear force.

Fortunately, in contrast to forces like gravity and the electrical attraction between charged particles, nuclear forces are short range, which means that two bits of matter cannot interact with each other via this force until they get very close together. Gravity is a weak force, but its effect extends indefinitely far away, so that the Sun can hold the Earth in orbit around it, and the mass of stars at the center of our galaxy many thousands of trillions of miles away can hold the Sun in orbit around it. Electrical forces are also long range and they are far stronger than gravity, but we do not normally notice them because matter is, on the whole, electrically neutral, so that the attractions and repulsions nearly balance and the strength of this force does not show itself. If matter were not so nearly neutral, electrical forces instead of gravitational would dominate our world.

How close must particles get before they interact via the strong nuclear force? The nuclear force decreases in strength very quickly at a distance of 0.00000000001 inches. This is a small fraction of the size of an atom – very short-range indeed by human standards. In order for nuclear reactions to occur, the particles involved must be brought together to these small separations. (To get some idea of how small the nucleus is, if we were to expand the diameter of an atom to one mile, the nucleus would be the size of a pea – and it contains 99.5% of the atom's mass.)

The problem is that the basic particle involved, the proton, is positively charged and positive charges repel each other. The force of repulsion between charges becomes larger and larger as they are brought closer and closer together. The amount of energy needed to overcome this repulsion and bring the nucleons together to such small separations turns out to be quite large, and only if the protons are thrown together quite violently can they overcome the electrical repulsion. This was a major technical obstacle in trying to produce the hydrogen bomb, which works by fusion, and it needed the explosive force of a fission bomb to crush the atoms of fuel close enough together to initiate fusion reactions; ordinary chemical explosives were not powerful enough.

The other way to bring the nucleons close together is by high temperature, since what we mean by "high temperature" is that the particles have very high average speed. When they collide at high speed, they come close together before the electrical repulsion drives them apart again. At high enough temperatures – several million degrees – the nucleons get close enough together for the nuclear force to act, and fusion can take place.

Quantum tunneling

Actually, the nuclear particles inside the Sun do not get close enough together to interact. Yet they do interact. The resolution of this apparent paradox goes to the heart of the strange and counterintuitive quantum theory.

Aside from jokes about misplacing one's glasses, ordinary objects in our world are either there or not there. If you place a cup on a saucer, you don't have to check whether it suddenly moves to the other side of the table, or whether the saucer decides to change places and appear on top of the cup. Macroscopic objects are well localized, and we are used to things being that way. This is one of the reasons that we have such a hard time picturing the world at the atomic and subatomic levels: our sensory apparatus did not develop with the need to deal with the (to us) bizarre behavior of matter on very small spatial scales.

What is the behavior that we would find so odd? Matter, it turns out, does not exist "at a place," but rather exists as a set of probabilities of being in a range of places. This is the famous wavelike nature of particles, and the usual analogy is made with a wave on a string, by pointing out that the entire wave has a certain size and it makes no sense to localize it to any smaller length than the distance between one wave and the next. This picture is not quite satisfactory, because we can look at an ocean wave or a vibrating string, and we can see the entire piece of string waving. We can point to each piece of it as being well localized; that is, we can see that the string or wave has parts and can identify each of the parts as it moves back and forth or up and down. In quantum physics, this is not possible. A small bit of matter really exists only as a cloud of probability until it interacts with another bit of matter (or radiation) at a particular place. Until the interaction occurs, the particle could be anywhere, although with decreasing probability as you search regions of space far from the most likely locations.

As with many of the other predictions of quantum mechanics, this one is hard to accept (Bohr remarked that if you are not profoundly shocked by the theory then you haven't understood it). What it means for our nuclear-burning problem is that, when two protons are thrown at each other, there is a small probability that they are actually in the same place, even if they cannot come all the way together under a classical (non-quantum) calculation. This effect was called "barrier penetration" – although these days it is often called "tunneling" – because the problem is formulated mathematically as one in which there is a barrier holding the two particles apart. Like a convict attempting to escape prison, the particle can tunnel under the barrier (metaphorically) and appear on the other side. It is even possible to calculate quite accurately the probability of tunneling, and such calculations were indeed carried out in the late 1920s and early 1930s.

The probability of a particle appearing on the other side of a barrier is very strongly dependent on the difference between the energy of the particle and the energy needed to penetrate the barrier. The energy of the particle is directly tied to the temperature in the center of the star, since the average energy per particle is what we mean by temperature, so that the penetration probability and therefore the rate at which the fusion reactions take place is very strongly dependent on the temperature. This is why small stars, with moderate temperatures in their cores, are so faint and live many billions of years: their fusion rate is relatively low. Large, massive stars, with high central temperatures, are exceedingly bright and burn up their fuel in only a few million years.

Formation of the elements

The fusion of light particles into larger clumps of so-called heavy nuclei is responsible for the formation of the elements. But a complete understanding of the processes involved in forming the elements took several decades, from early work by George Gamow on nuclear fusion in 1928 until the late 1950s. It seems that there are three major nuclear cooking-pots in which element-building takes place; two of the three directly involve stars:

1. The first three minutes. During the first few minutes of the Big Bang, the temperature and density of the expanding fireball were high enough for the fusion of protons and neutrons into heavier elements to occur, but cool enough (less than a billion degrees) that the heavier nuclei were not torn apart as soon as they formed. This happened at the age of about one minute; by the age of three minutes, the universe had expanded and cooled enough so that the formation of heavy elements stopped.

The basic process was worked out by Ralph Alpher in 1948, in collaboration with George Gamow (and a ghostly Hans Bethe, added by Gamow as a joke to make the authors be $\alpha - \beta - \gamma$). Only the lightest elements formed this way: hydrogen (1 proton), deuterium (1 proton and 1 neutron), some tritium (1 p and 2 n), a large amount of helium (2 p and 2 n), a small amount of light helium (2 p and 1 n), and very small amounts of lithium (3 p) and beryllium (4 p), formed by the combination of deuterium and tritium. The problem with getting elements beyond helium in any great numbers is that the next nucleus, having a mass of 5, is unstable and doesn't live long enough for a sixth nucleon to join in, except in rare multiparticle interactions. So the heavier elements must have been formed after the Big Bang.

2. Stellar interiors. Largely through the work of Hans Bethe in 1938, the fusion of light nuclei into heavier ones in the cores of stars was understood. With this development, the energy source of stars was explained, and a first step was taken in explaining the formation of some heavy elements out of lighter ones. The basic processes, taking protons into alpha particles (helium nuclei) and eventually up to carbon, were understood. (A temporary state involving 3 alpha particles, allowing it to form carbon-12, was discovered by Fred Hoyle, allowing the Universe with all of the elements we are used to.)

3. Novae and supernovae. In 1957, a massive survey of many possible fusion processes was carried out by Burbidge, Burbidge, Fowler, and Hoyle (and independently by Cameron). This included processes not occurring in ordinary stellar interiors, such as the rapid (r-process) and slow (s-process) neutron fusion reactions, which were later understood to occur in nova and supernova explosions. In this way, the heavy elements, through uranium and the even heavier trans-uranic nuclei, could be produced in these massive stellar explosions. Many of these heavy nuclei decay very rapidly, or at least rapidly compared to the age of the universe, and are not normally found in nature at the present time.

So it is true that we are stardust: most of the elements that make up our bodies, and nearly everything we see around us, were manufactured inside stars.

THE EVOLUTION OF SOLAR ACTIVITY

Though you can't usually look at the Sun directly, sometimes at sunset it is dimmed just enough by distant haze that you can see what is on it. Occasionally, a small dark spot can be seen, even with the naked eye.

These dark regions, called sunspots, were studied in 1610 by Galileo, who turned the newly invented telescope on objects in the sky (Fig. 2.3). About the same time, Christopher Scheiner, a Jesuit astronomer, saw the sunspots, and a lively debate over their origin and significance ensued. We now know that the sunspots are relatively cool regions on the Sun, over 1000

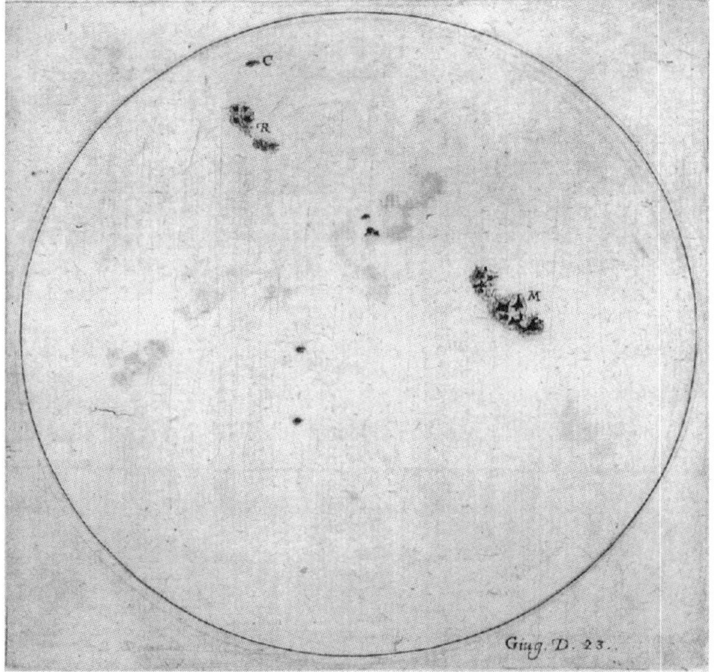

FIGURE 2.3. Drawing of sunspots made by Galileo in 1612. Using sequences of such drawings he showed that these blemishes were on the surface of the Sun, which was therefore not "perfect."

kelvins cooler than the surrounding solar surface – still hot, by terrestrial standards, of course. As a consequence, the sunspot gas is less bright than the surrounding gas, appearing dark in comparison. The number of spots on the Sun increases and decreases from year to year, with an average of about 11 years from one maximum to the next; this regular variation is called the sunspot cycle.

Early in the 20th century, George Ellery Hale examined the spectra of sunspots. He discovered that certain spectral lines were split into three parts: instead of seeing just a single sharp line in the spectrum, one sees the expected line plus an additional line on either side of it. Such splitting had been discovered earlier in a laboratory setting by the Dutch physicist P. Zeeman. It is due to the presence of a strong magnetic field, which alters the energy levels in some atoms, so that the atomic transitions which absorb and emit light are slightly changed in energy. The magnetic field shifts the line both higher and lower in energy, so that the splitting can show up on both sides of the original line. Stronger magnetic fields result in greater splitting of the levels. In the Sun, Hale measured the splitting in sunspots as due to a magnetic field of 3000 gauss, about 3000 times higher than the 1 gauss average magnetic field on the Earth (Fig. 2.4).

The strong magnetic fields in sunspots limit the conduction of heat flowing up from the interior of the Sun, diverting the upflow from the interior and keeping the spots relatively cool. We can even detect, from their spectra, several molecules in sunspots, while the normal solar surface is so hot that molecules are broken apart.

About a hundred years ago, Richard Carrington in England noticed that at the start of a sunspot cycle, sunspots initially emerge at high latitudes from the solar equator. Then, as the years go by, new spots form closer and closer to the equator. When they get within about 15 degrees of the equator, a new set of sunspots starts forming again at higher latitudes. He plotted

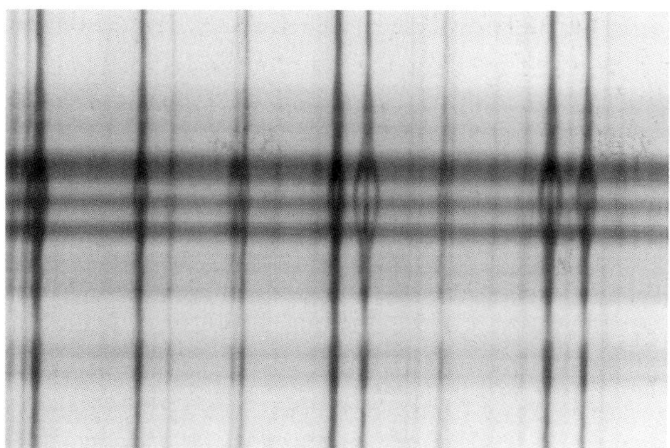

FIGURE 2.4. Measuring the magnetic field in a sunspot via the Zeeman effect. The strong spot field splits the line into three components.

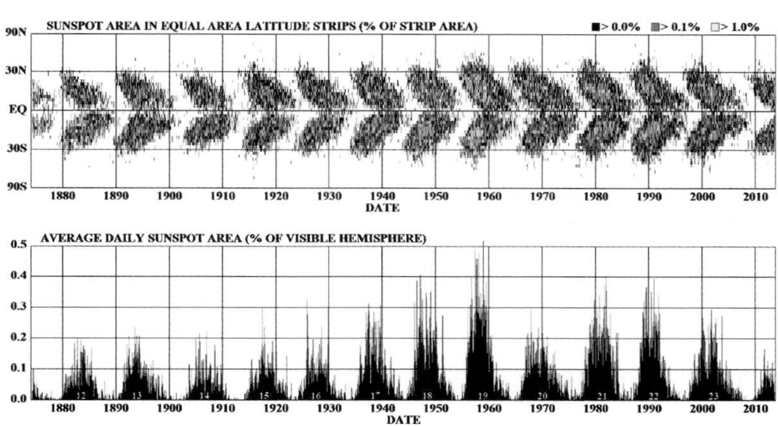

FIGURE 2.5. The latitude of sunspots appearing on the Sun and the percentage of the visible solar disk covered are plotted against time for the most recent solar-activity cycles.

the effect in what has become known, for obvious reasons, as the "butterfly diagram" (Fig. 2.5). You can see that as the 11-year cycle continues (from left to right), sunspots first appear at a high latitude and then appear at lower latitudes. The appearance of new spots at high latitude marks the beginning of a new cycle.

Scientists are unable at present to predict how strong a cycle will be in terms of sunspot number. For example, a set of papers in the late 1990s tried to predict how high the peak of the cycle would be in the year 2000. Of the set, one paper predicted that it would be higher than average, another predicted that it would be lower than average, and a third predicted that it would be average. (It turned out to be about average relative to recent cycles.) It has also been suggested that there are other cycles with longer periods than 11 years that are superimposed on the 11-year cycle. The evidence for this is still marginal, and unless we can find a way to extend the historical sunspot record back before 1610 we will just have to wait longer to obtain higher statistical certainty for such cycles.

In the meantime, the 11-year sunspot cycle is very obvious. Later in this book, we will see that the sunspot cycle is only a symptom of a more general solar-activity cycle. What drives the sunspot cycle is a cycle in the intensity of the Sun's magnetic field strength. These magnetic fields, generated inside the Sun, create a large assortment of dynamic effects which are collectively called the solar activity cycle. The sunspot cycle is the most easily observed manifestation of the activity cycle.

In the 19th century, Edward Maunder noticed something puzzling about the historical sunspot record: there seemed to have been very few spots on the Sun in the decades after Galileo's discovery. Sunspots were generally absent for about 70 years, from about 1645 to 1715. Maunder's work was forgotten for decades and then rediscovered by John Eddy of the High Altitude Observatory in the 1970s. This "Maunder minimum" may indicate that the sunspot cycle turned off for a time in the 1600s, which would imply that the sunspot cycle is superficial rather than fundamental to the Sun. Long-period cycles like

FIGURE 2.6. Early 17th century painting by Hendrick Avercamp (1585-1634) showing ice skating on a canal in Holland during the Maunder Minimum. The connection between the Maunder Minimum and the Little Ice Age is unclear and is currently being debated.

this may turn out to be important for understanding the linkage between the Sun and with weather on Earth. In particular, the Maunder minimum corresponded to a relatively cool time in Europe sometimes known as the Little Ice Age (Fig. 2.6). The Southwest U.S. was in drought at the same time.

Were there really fewer sunspots during 1645–1715, as Eddy suggests? Or were people just not watching closely enough? The fact that individual papers appeared reporting on sunspots indicates that they were indeed a big thing, and would have been reported. Indeed, reports at the time state that observers had been searching in vain for years to see even a single spot on the Sun. Further, other indicators of the Sun's magnetic field also were weak or absent during much of that time. For example, records show that there were few auroras and that the corona of the Sun was not strong when seen at eclipses during that

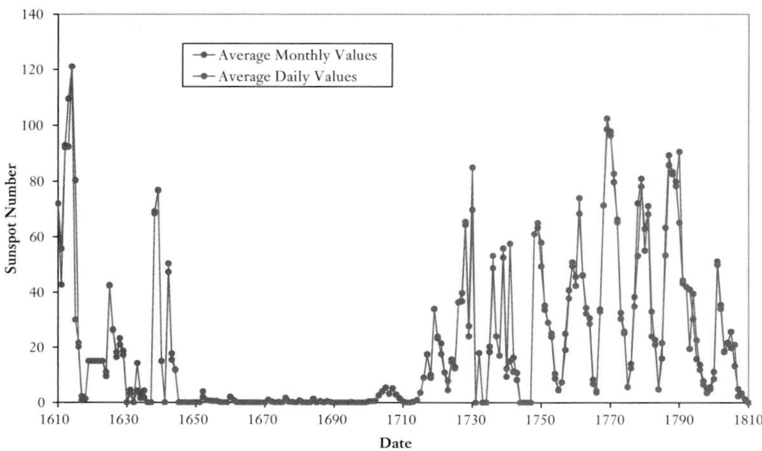

FIGURE 2.7. A recent reconstruction of sunspot numbers during the 17th and 18th centuries shows clearly that the number was close to zero for an extended period of time during the Maunder Minimum.

interval. Recent work has included many contemporary publications from the time to construct an accurate record of sunspot number during those years and finds strong evidence that the sunspot cycle had essentially vanished (Fig. 2.7).

In the 1980s, work started by Olin Wilson and others at the Mt. Wilson Observatory established that several other stars also have solar-activity cycles. By following a certain spectral feature that is typical of solar activity yet is visible from the ground, it is noticed that the strength of this feature varies in nearby solar-type stars with periods ranging from a few years to 15 years or so. Some stars may have longer periods than that, but more years of observation are needed to be sure of that. They also found some solar-type stars that appeared to be in a Maunder minimum phase of very low activity. Enough observations of this type could establish how much of the time the Sun is in a protracted minimum of activity, by determining what fraction of the solar-like stars are in such a state at any given time. This work is a good example of the solar-stellar connection, the use of solar and stellar data to complement each other: our knowledge of the Sun helps us to interpret the stellar data, and the stellar observations help us to test theories that are developed about the Sun.

THE SUN TODAY

The Sun is in the prime of its life, about halfway through its ten billion year history. We expect slow changes in the Sun during this time, especially as the hydrogen fuel in its core is gradually converted to helium and heavier elements. Such slow changes are undetectable on human timescales, but we also know that there is the very rapid 11-year sunspot cycle, and perhaps other cycles as well.

There are indications that such short-term (compared to the 10 billion-year lifetime) variations in the Sun show up as small changes in the Earth's climate or weather patterns. So we may naturally wonder whether they show up in the total energy output from the Sun. In Chapter 7 of this book, we will see how the American scientist Charles Greeley Abbot tried for decades to ascertain if the sun shines steadily. The accuracy necessary was beyond his capabilities, but with the launch of space satellites, we began to be able to make this measurement.

The solar constant is the amount of energy that reaches each square meter at the radius of the Earth's orbit. It is taken to be the amount that reaches the top of our atmosphere, so that we can measure it without regard for the vagaries of atmospheric effects. Starting with the Solar Maximum Mission in the 1980s, a series of satellites have been launched, with instruments able to take in and measure all of the solar energy, unfiltered by terrestrial atmosphere. The instrument used – essentially a cavity that lets radiation in and then measures the resulting temperature of the cavity – improved the ground-based accuracy of 1.5 per cent

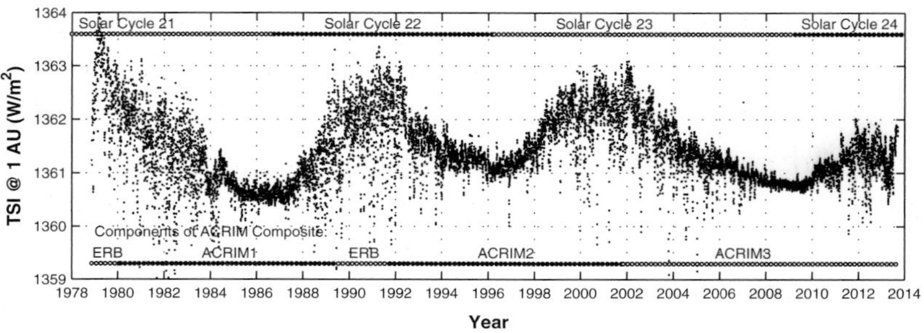

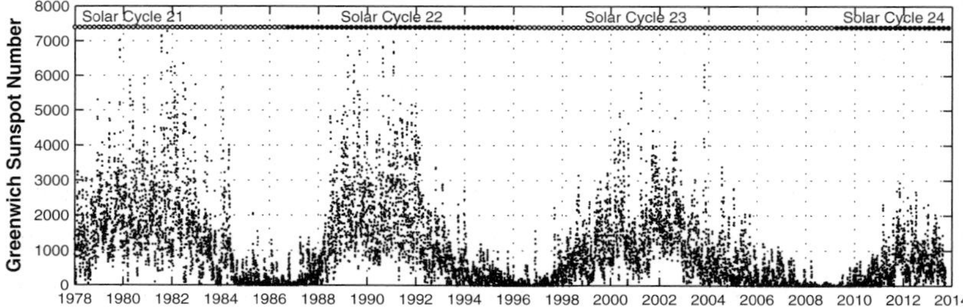

FIGURE 2.8. Measurement of the Total Solar Irradiance (TSI; formerly, before it was found to be varying, the "solar constant") over three sunspot cycles from several different satellites. The variation is by about 0.1%. NASA's Glory spacecraft, which was to have resolved calibration differences among the several satellites measuring TSI, failed at its 2011 launch.

to the current 0.1 per cent. The result is about 1366 watts per square meter (Fig. 2.8).

But the spacecraft also found deviations of about 0.2%, often coinciding with the passage of a large sunspot, or spot group, across the Sun. Only with that measurement did it become known that sunspots *per se* actually diminish the solar radiation, while nearby parts of the surface become a bit brighter. Frightening for a time was the discovery that the solar constant was diminishing at a rate that would be serious for us on Earth if it continued for only a few decades, very short on an astronomical timescale. Fortunately, when measurements were made over enough years, we saw that the overall change in the solar constant was linked to the solar activity cycle. When the sunspot number started rising, the solar constant started rising, until it reached the value seen at the previous peak of the cycle. Several satellites, with names like Upper Atmospheric Research Satellite and Earth Radiation Budget Satellite, and ACRIMSAT (Active Cavity Radiometer Irradiance Monitor Satellite) have continued to monitor the solar constant – which isn't really a constant at all, and whose changes may be linked to changes in the Earth's weather patterns, as we'll discuss in Chapter 7.

The sunspot number continues to be counted. It is tabulated by the Solar Influences Data Analysis Center in Belgium, which averages the reports of dozens of astronomical sites. Another tabulation is carried out by the American Association of Variable Star Observers in the United States, which averages reports made by many amateur astronomers. These counts are available on the Internet each month at http://sidc.oma.be (Fig. 2.9), as

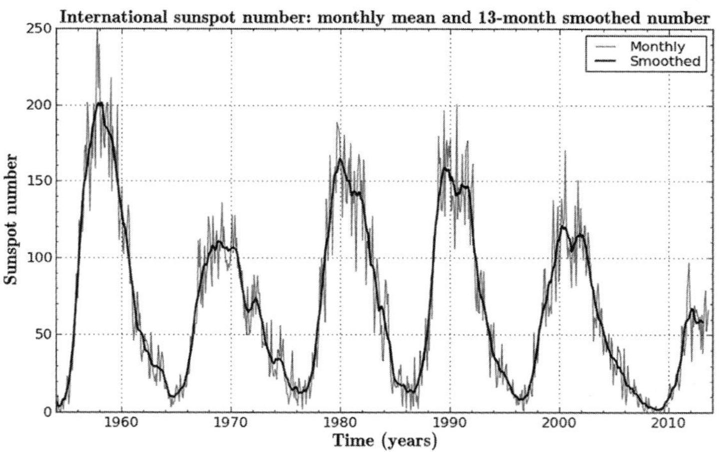

FIGURE 2.9. The monthly averaged sunspot number, with the smoothed values shown in a darker line, for the five and a half recent solar cycles, from the Solar Influences Data Analysis Center at the Royal Observatory of Belgium.

is the butterfly diagram. The "sunspot number" has a historical significance, since it has been kept for about 150 years and can therefore be used to extend the range of certain studies back in time. But it isn't really the number of sunspots, which after all come in a wide range of sizes. The mere number of spots may not be as significant as, say, the total area of the solar surface covered by spots.

But measuring the area covered by spots is very difficult and tedious, so that a compromise measurement is used. The "sunspot number" is defined as ten times the number of groups of spots on the Sun plus the number of individual spots, all multiplied by a "personal constant." The personal constant takes account of each individual observer's tendency to be biased in a consistent way, such as how he or she decides whether there are one or two spots in an ambiguous region, and for such tendencies to vary from one individual to another.

Let us consider that there is one spot on the side of the Sun facing us. (The sunspot number does not take account at all of the fact that from Earth we are seeing only half the Sun at a time; spots on the other side simply are not part of the count, even though the STEREO satellites and helioseismology now allow us to detect spots on the solar farside.) If there is one sunspot present, then the sunspot number is 1 group + 1 spot and the sunspot number is 10 + 1 = 11. If there are two sunspots present, it matters how they are distributed. If they are together, then there is 1 group + 2 spots and the sunspot number is 10 + 2 = 12. If they are separate, though, then there are 2 groups + 2 spots and the sunspot number is 20 + 2 = 22. The sunspot number, as you saw in the figure, can be as low as 0 and, averaged over a month, can exceed 180. For an individual day, the sunspot number can reach 250 or greater.

The butterfly diagram was plotted showing the average daily area of the Sun covered with sunspots, calculated as a percentage of the entire solar disk facing us. A typical sunspot covers about

0.02% of the disk, so that at most half a per cent of the Sun might be covered. More typically, only about a tenth of a per cent is occupied by spots. If our Sun were at the distance of other stars, we would not detect this activity, whereas we do actually see noticeable spot-related brightness changes on several nearby stars. Thus, these stars must have much more coverage by spots, and our Sun seems to be relatively low in activity compared to its neighbors.

THE FUTURE

The Sun is reasonably steady now, but starting in about three billion years, changes in its core will be reflected in a change at the surface. The outer atmosphere of the Sun will begin to swell so much that our own atmosphere will heat up, forcing us to leave Earth. We on Earth would eventually be incinerated, if we stayed. Fortunately, we have billions of years to arrange to go elsewhere.

Eventually, the Sun's outer layers will brighten and redden faster and faster. These outer layers will envelop even the current orbits of Mercury and Venus, reaching approximately the current orbit of the Earth. The Sun will then be a type of star known as a red giant. We know of many such stars in the sky. It will be red because the outer layers will be relatively cool, a bit more than half the temperature they are today. But even though the solar surface will be cool, it will be so close to Earth that oceans will boil and humans could not survive. The solar wind, an outflow of particles, will be so strong that the Sun will be losing mass rapidly, which will make the planets recede a bit, but not enough to save us.

While the outer layers of the Sun are enlarging, the core of the Sun will be collapsing inward, which will raise the central temperature and pressure. About five billion years from now, the core will rise from its current 15 million degrees up to about 100 million degrees. At that point, helium atoms will begin to fuse together three at a time, first making beryllium, then carbon. This is known as the triple-alpha process and it produces what is known as the helium flash, because the process commences very rapidly and exhausts itself in such a short time interval, as short as a few seconds for some stars. The energy from the helium flash will expand the Sun's core. Because it then gives off less energy for a while after the flash ends, the outer layers shrink again. At this time, helium will fuse into carbon in the core in a steady fashion. Outside the resulting core of carbon, shells of fusing helium and, further out, fusing hydrogen, will exist. This stage will last a few hundred million years.

The shells of fusing hydrogen and helium burn unstably, causing pulsations that continually increase in strength, and the stellar wind increases by a billion times, carrying out large quantities of material into interstellar space. The material carried out, whether by pulsing or by the unstable solar wind, will drift away from the Sun. As the outflowing wind increases in strength and speed, new material catches up with the old, piling up the material in shells around the Sun.

We see hundreds of such stars in the sky. They are known as planetary nebulae, though the word "planetary" means only that a hundred or so years ago they appeared in the telescopes of their time as fuzzy disks like the planets Uranus and Neptune. Actually, they are the clouds of gas around dying stars that were once like our Sun. Images taken by the Hubble Space Telescope show the many different forms that planetary nebulae can take, depending on the details of the irregular pulsing of the gas and on whether or not there are directional jets of gas present, aligned to the stars' original spin axes (Plate I). The Ring Nebula in the constellation Lyra is a planetary nebula familiar to amateur astronomers, since it is visible in even small telescopes.

The planetary nebula stage of the Sun's future will last only about 1000 years, a snap of a finger in the Sun's 10 billion year overall lifetime. After that time, the gas surrounding the star will expand and cool so much that it will become invisible. It will carry the carbon and other elements out into space, providing elements necessary for life forms such as ourselves.

The core of this remnant Sun will cool and contract, until the electrons inside it cannot be pushed together any more by the gravity of the remnant's mass. Our Sun will then be an Earth-sized star, of the type known as a white dwarf, and will end its life in this way, slowly cooling until the end of time.

SUGGESTIONS FOR FURTHER READING

Brody, Judit, *The Enigma of Sunspots: A Story of Discovery and Scientific Revolution* (Floris Books, 2003).

Impey, Chris, Lunine, Jonathan, and Funes, José, *Frontiers of Astrobiology* (Cambridge University Press, 2012).

Pasachoff, Jay M. and Filippenko, Alex, *The Cosmos: Astronomy in the New Millennium*, 4th edition (Cambridge University Press, New York, 2014).

Phillips, Kenneth J. H., *Guide to the Sun* (Cambridge University Press, 1992).

Sagan, Carl, and Mullen, George, "Earth and Mars: Evolution of atmospheres and surface temperatures," *Science*, **177**, 52, 1972.

Zirker, J.B., *The Magnetic Universe: The Elusive Traces of an Invisible Force* (Johns Hopkins University Press, 2009).

3

What We See: The Solar Disk

The Sun is basically a hot ball of gas, powered by nuclear reactions at its core and eventually radiating that power out into space at its outer surface. The temperature at the core is millions of degrees, so that the "light" produced there is mainly x-rays. But this light must travel through an enormous amount of matter to get to the surface of the Sun, being scattered, absorbed, and re-emitted so many times that an astounding 100,000 years and possibly ten times more are needed for the energy generated in the core to get to the surface.

This energy is spread over a larger and larger area as it moves outward. The average temperature of the radiation drops as it moves out, but in such a way that the total – that is, the local intensity multiplied by the area of the larger and larger surface – remains a constant. By the time the energy reaches the visible surface of the Sun, the temperature has dropped to about 5800 K (10,000°F). This puts the radiation emitted into the visible part of the spectrum – our eyes almost certainly having adapted to the available wavelengths, thereby making them the "visible" part.

THE PHOTOSPHERE

Every day that is not cloudy, we see a bright disk of light in the sky. Visually, the Sun seems to have a surface, and the light we see comes from that surface. "Photos" is the Greek word for "light," and "sphere" accounts for the spherical shape of the solar surface, so astronomers use the term "photosphere" to mean that visible surface of light.

But if the Sun is made of gas through and through, how can there be anything that we call the Sun's surface? And why does the Sun seem to have a sharp edge? To understand the answers, we need to know some of the properties of gases. Most gases are transparent, but this seeming transparency is only because we are looking through them for small distances. Imagine a smoke-filled room, for example (a location that does not exist in quantity any more): we can usually see our hands at the ends of our arms, but may not be able to see the far wall. The smoke is opaque enough that we cannot see through 20 feet of it.

We measure how opaque something is with a quantity named "opacity," the opposite of transparency. Something that has opacity zero is completely transparent. We can see only murkily through something that has opacity 1. And by the time the opacity is 5 or more, we effectively can't see through it at all. (Technically, when light goes through a gas of opacity 1, its intensity is diminished by a factor of e, a natural constant equal to 2.71828.... Opacity 2 is a factor of e^2 or roughly 7; and so on. Opacity 5 is a factor of 148, so less than 1 per cent of the light gets through.)

So when we look at the center of the Sun's disk, we see into the gas until the opacity builds up enough so that light from a lower level effectively doesn't get out. It turns out that most of the light we receive is from a level at which the opacity is 2/3. We get a little light from farther out, and a little light from farther in, but it is convenient to talk of the level at which the opacity is 2/3 as the surface of the Sun. The photosphere is a few thousand miles thick, but we define the level at which the opacity is 2/3 as the base of the photosphere.

The edge of the Sun

When you can see the disk of the Sun – perhaps dimmed by fog or haze at sunset, or when you look through a suitable, safe solar filter – its edge seems to be sharp. How sharp is it?

In the case of the edge of the Sun, it doesn't have to be actually sharp; it only has to be sharp enough to seem that way to the human eye. The pupil in your eye through which the light comes is only a few millimeters across – sometimes only a millimeter or two in bright light and rarely more than 8 millimeters. The sharpness of an image in an optical system is limited by its size because of diffraction, with larger optics being needed for higher resolution. Your eye can resolve only about a minute of arc, an angular measure that is about 1/30 the diameter of the Sun or Moon.

The question to ask is: what is the angle between a line of sight that hits the disk of the Sun (the photosphere) and a line of sight that seems to go past the Sun into space farther out, as shown in Figure 3.1? That angle turns out to be much less than 1 minute of arc, and so isn't resolved by the human eye. Thus the edge of the Sun – the edge of the photosphere – is narrow enough that it appears sharp to us. In reality, the edge is

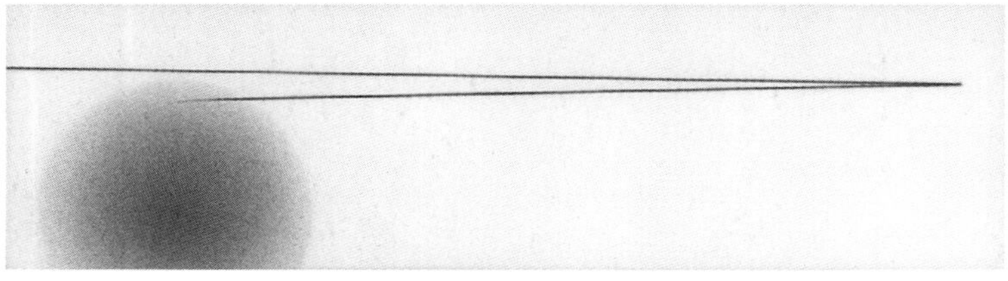

FIGURE 3.1. The edge of the Sun, and why it seems sharp to the unaided eye.

"fuzzy" by about 100 miles, which is large by humans standards, but small on an object as large as the Sun. The same effect also makes clouds in the Earth's atmosphere appear to have sharp edges.

Limb darkening

On a photograph of the photosphere, you may also notice that the Sun does not appear uniform in brightness across its disk: the parts of the disk near the edge appear somewhat darker. Since the edge of the Sun (or moon or star) is called its limb, the phenomenon is known to astronomers as limb darkening (Fig. 3.2).

You can see the effect if you use one side of a pair of binoculars to project a solar image onto a surface. (Be careful not to look up at the Sun through the binoculars; use the binoculars

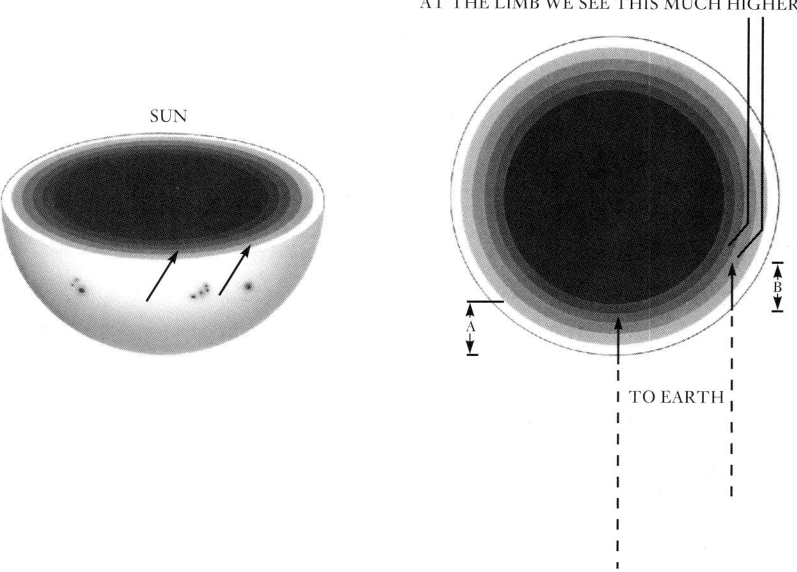

FIGURE 3.2. Limb darkening – visibility into the Sun near the limb stops at higher, cooler and therefore darker levels of the atmosphere.

only to project the solar image onto a screen.) The image will show sunspots and, near the edge of the Sun's surface, will look darker.

We understand the phenomenon of limb darkening in terms of the idea above, that the Sun is a murky gas. When we look toward the center of the Sun, we see in until our line-of-sight penetrates enough material to make the opacity add up to about 2/3, that is, for the gas to become effectively opaque. But when we look near the Sun's limb, we are looking diagonally through the Sun's gas and have to penetrate more material to see down to a given depth below the surface. This means that the opacity adds up to about 2/3 at a higher level, farther from the Sun's center. So when we look near the limb, we are seeing photospheric gas from a higher atmospheric level than when we look near the center of the Sun's disk.

Why is the limb darker? Because the upper level of the Sun's atmosphere is cooler than the deeper level, so the light emitted from the higher levels is less intense. What we learn from studying the limb darkening in the visible part of the spectrum, is that from the base of the photosphere upward for a few thousand miles, the Sun's atmosphere is getting cooler.

In other parts of the spectrum, the same effect occurs, in that near the Sun's limb we are getting radiation from higher, cooler levels. But in some parts of the spectrum, such as the millimeter-wavelength radio radiation, our instruments detect a limb brightening instead of a limb darkening. This is a first indication that something other than a decrease of temperature with height happens above the photosphere, and we will have much to say about this in the rest of this book.

The temperature of the Sun's surface

The Sun gives off what we call "white light": all the colors of the rainbow. Our eyes naturally adapt to sunlight, so we see a white sheet of paper as white no matter whether it is midday, with the Sun overhead, or at sunset, with the sunlight that filters through the atmosphere becoming reddish. But color film is less forgiving – photographers often find their sunset pictures looking excessively reddish, or they use a bluish filter to remove some of that annoying reddish tinge. So we know that sunlight contains red in it, and of course the blue of the sky comes from blue wavelengths that were removed from the incoming sunlight while letting the reddened portion through.

Visible light, the light we can see with our eyes, ranges in wavelength from about 400 nanometers (400 nm) to 700 nanometers (700 nm), where 1000 nanometers is 1 micrometer, which is 1/1000 of a millimeter. [In the metric system, milli- (a prefixed m) means 1/1000; micro- (whose symbol is a Greek mu, μ) means 1 millionth; and nano- (a prefixed n) means 1 billionth; m written by itself is the symbol for a meter.] We are talking, in customary American units, of about a millionth of a yard for a wavelength of light.

The distribution of the Sun's energy looks like a bell-shaped curve with a peak in the yellow-green region of the spectrum. Such a curve was shown in Fraunhofer's historical measurement of the distribution of the energy from the Sun, Figure 1.1. Intensity distribution curves of this sort for theoretically perfect hot, radiating bodies at different temperatures are known as blackbody curves. Depending on the temperature, their peaks fall at different wavelengths. Blackbody curves that resemble the distribution of the Sun's energy come from gases of about 5800 kelvins (to convert to degrees Celsius, subtract 273, giving about 5500 degrees Celsius), corresponding to about 10,000 degrees Fahrenheit. Note that the temperatures are usually given only to one or two significant digits, with zeros following; we often don't know them more accurately than that, and adding extra non-zero digits merely gives a false impression of accuracy.

A blackbody is a theoretical construct, an object that doesn't really exist but which would give off a standard distribution of radiation. The Sun is surely not really a blackbody; after all, it changes in temperature from place to place, both inward and outward and from side to side through its atmosphere, whereas a blackbody is, by definition, at one specific temperature. But in visible light, the radiation from the photosphere is close to that of a blackbody. This closeness allows us to assign a temperature to the level of the solar photosphere from which most of the Sun's light comes.

The spectrum of the Sun's surface

The overall distribution of the continuous band of colors in the Sun's spectrum can be used to find the photosphere's temperature, as we just saw. But examination of the solar spectrum in great detail gives much more information.

The Sun's spectrum was first described in modern terms by William Wollaston in 1804, who named a few colors in the spectrum. But technology advanced soon thereafter, and Josef Fraunhofer, in Germany, in 1814 provided the spectrum that we saw in Figure 1.1. Fraunhofer noticed that the continuous rainbow of the Sun's radiation is crossed by dark lines in which the strength of the radiation is greatly diminished. These regions are known as absorption lines or, usually, Fraunhofer lines (Fig. 3.3).

The spectrum is really the image of a slab of incoming sunlight spread out from side to side according to wavelength. The slab of light is produced by a slit in a metal surface on which the Sun's image is allowed to fall from a telescope. The light that passes through the slit enters a device called a spectrograph, which spreads out the spectrum ("spectro-") and records it ("-graph"). The Sun is so bright that its light can be spread out to a very great degree, while still leaving the intensity high enough to be recorded. Figure 3.4 shows a solar spectrum so

FIGURE 3.3. Fraunhofer's original spectrum shown on a German postage stamp. The spectrum is crossed by the dark absorption lines that Fraunhofer discovered in the Sun's spectrum.

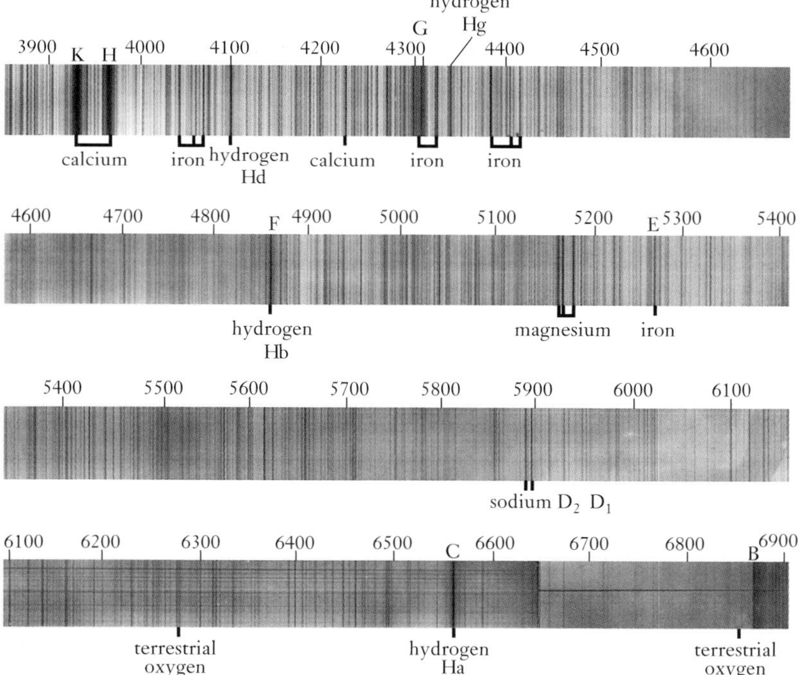

FIGURE 3.4. Solar spectrum with some prominent spectral lines identified.

long that it is folded into several strips, each covering only a few nanometers of wavelength. The most detailed of such observations have been used to measure about a million Fraunhofer lines. All other stars also have absorption lines, though they are not bright enough for us to spread out their spectra as far, and not as many lines are known. Millions of distant stars, though, seem to have spectra just like that of the Sun. These stars are known as type G2 dwarfs, and the Sun is quite typical of the class.

To understand what causes the dark Fraunhofer lines, we must consider what happens when light passes through a gas. The key process is excitation or ionization of the atoms of the gas by the light coming through it from below. When the overlying gas is cooler than the source of the light (which is the underlying gas, which emits a continuous broad range of wavelengths), the energy from the light can be absorbed by the atoms in the cooler gas. These atoms take up the energy at only a few particular wavelengths for each type of atom.

Energy is neither created nor destroyed in the atoms, but the energy removed from going straight ahead is radiated by the atom in all directions. Thus less of the energy, at a few specific wavelengths out of the broad continuum of wavelengths, is left to go straight ahead. When we look back through the cooler gas at the hotter source, we see a set of dark lines subtracted out of the continuum light. The set of wavelengths taken up for each type of atom at a given temperature acts like a fingerprint in allowing scientists to identify what atoms are present and how hot they are.

In the Sun, the fact that we see Fraunhofer lines tell us that we are looking through a cooler gas at a hotter one. This conclusion is the same we reached from the study of limb darkening. Indeed, the temperature that we measure by identifying the Fraunhofer lines is roughly the same as that we get from the continuous spectrum. Differences in detail come from the fact that the Sun is neither homogeneous nor a true blackbody.

Some of the most prominent Fraunhofer lines that we see in the solar spectrum are from hydrogen. It might seem that the hydrogen lines should be especially prominent since 90 per cent of the gas in the Sun (and in the Universe in general) is hydrogen. But it turns out that the strongest hydrogen Fraunhofer lines are at ultraviolet wavelengths too short for us to see in visible light. The Fraunhofer lines we see in the visible are secondary lines, still strong enough to be prominent but not strong enough to be very prominent. Indeed, when Cecilia Payne (later Payne-Gaposchkin) suggested, at Harvard in 1929, that hydrogen was the dominant gas in the Sun and stars, her conclusion was quickly rejected by the then most prominent astronomers of the time. It took years before the world in general came around to her conclusion. She became used to being a pioneer – she was, much later, the first female professor in all of the Faculty of Arts and Sciences at Harvard University.

What is the hydrogen spectrum like in the visible? Since hydrogen is the simplest element, having just one electron, its spectrum is also fairly simple. It is a series of lines, starting with a bright red line, then a less bright blue line, and then other lines still less bright in a sequence converging at shorter and shorter wavelengths through the blue and violet (Fig. 3.5). These lines were analyzed in the 19th century by the Swiss schoolteacher Johannes Balmer, and are still known as Balmer lines.

The strongest Balmer line, the red line, is known as H-alpha. The other lines are lettered, continuing in the Greek alphabet: H-beta, H-gamma, and so on. Whenever you spot that sequence of converging absorption lines, you know that hydrogen is present. By analyzing the strengths of the lines, you can often tell how much hydrogen is present and how hot the gas is. Balmer realized that each state of the hydrogen atom at a certain temperature corresponded to a certain amount of

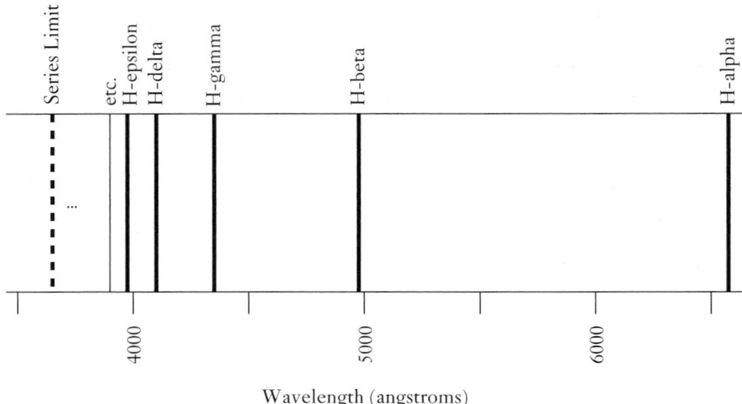

FIGURE 3.5. The spectrum of hydrogen at visible wavelengths, showing the first five lines in the series named after Balmer.

energy, and that the series of lines we now know as Balmer lines corresponded to changes in the state of the atom from one well-defined energy to another. We now know that these changes come from electrons jumping from one energy level in the atom to another.

The strongest Fraunhofer lines in the visible spectrum are a pair of purple absorption lines, just into the ultraviolet at the limit of our eyes' visibility. Fraunhofer had labelled the strongest lines in his spectrum: A, B, C (which turns out to be H-alpha), and so on, and one of these strongest lines is Fraunhofer's H-line. Fraunhofer had used the letter I to indicate the end of his spectrum, but a less-known scientist later in the 19th century had associated the letter K with the second of the two strong ultraviolet lines. So these "H and K lines" are the pair of very strong absorption features in the violet spectrum of the Sun. They turn out to be from once-ionized calcium, calcium that has lost one electron. They are the fundamental lines that are absorbed by a calcium atom, which makes them so strong, given that calcium is a reasonably abundant element on the Sun.

They are less strong than the fundamental hydrogen lines that we can't see in the ultraviolet, though stronger than the Balmer series lines of hydrogen that we see in the visible.

Most of the lines that appear in the Sun's spectrum are from iron and from similar elements. Iron just has so many electrons (26 in the un-ionized state) that the spectrum resulting from the various jumping around of electrons, so simple for hydrogen, is very complex in iron. None of the individual iron lines is exceptionally strong, though. Still, some have special uses, paricularly those that are sensitive to magnetic fields (Fig. 3.6).

Historically, people spoke of the Sun's "reversing layer": hot gas at the base of the photosphere and cooler gas above it causing the Fraunhofer lines at higher levels. We now know that the situation is not quite so simple. The Sun's continuous radiation and the formation of the Fraunhofer lines are mixed through the photospheric layers, with both forming gradually with height and with many details depending on exactly which line you are studying. The overall temperature of the photosphere seems to range from 6,400 kelvins (11,000 degrees Fahrenheit) at the lowest levels to 4,400 kelvins or so (7,500 degrees Fahrenheit) at the upper heights.

The composition of the solar photosphere

From the Fraunhofer spectrum, astronomers can analyze how much of each element is present. They have found almost all the known chemical elements there. The quantity of an element is known as its abundance, and most abundances astronomically are compared with that of hydrogen or with that of carbon. Normally these abundances are given by number, and not by mass. For example, if we have 9 atoms of hydrogen and 1 of helium, hydrogen is 90% of the elements by number. But the hydrogen atomic mass is 1 and the helium atomic mass is 4, so the total atomic mass of the atoms is $9 + 4 = 13$. The abundance of

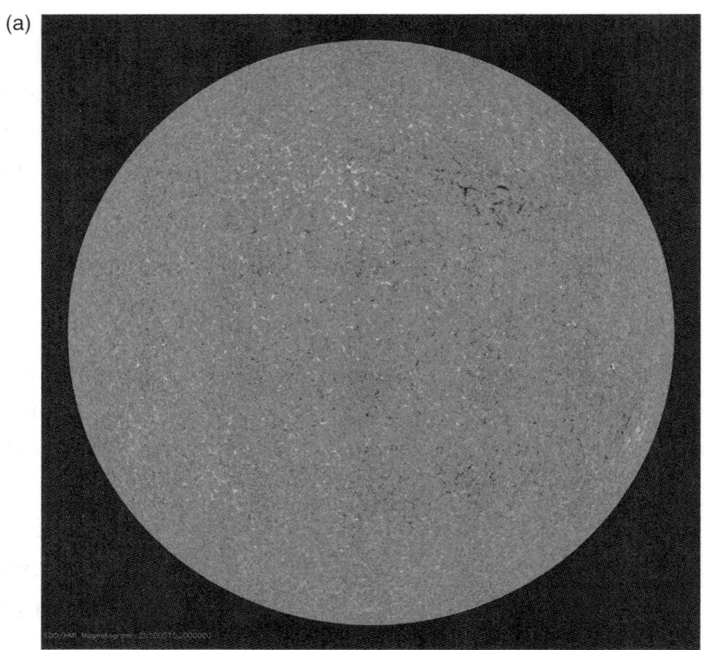

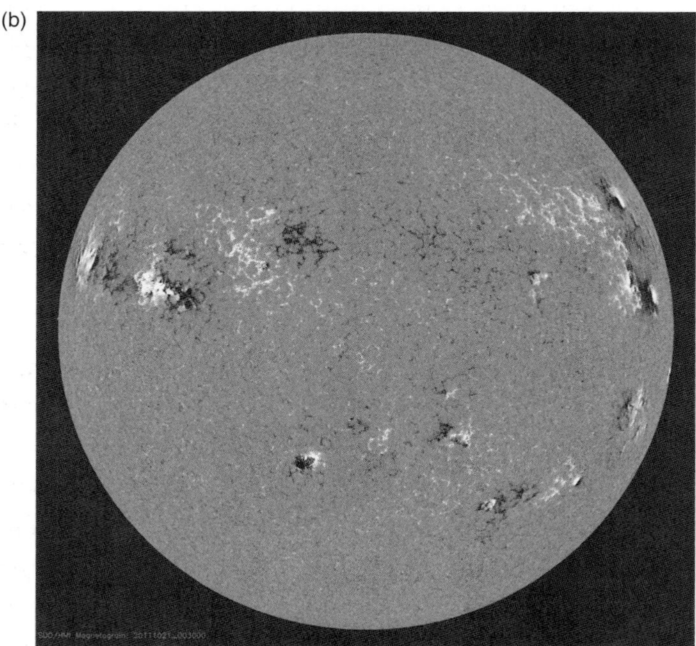

FIGURE 3.6. Surface magnetic fields on the Sun near solar minimum in May, 2010 (a) and near maximum in November, 2011 (b).

hydrogen by mass, if we choose to compute it that way, is thus 9/13 or about 70 per cent. (This also happens in situations closer to home. For instance, the alcohol content of wine is usually given by volume, rather than by weight. Since alcohol is lighter than water, the amount would seem lower if it were stated by weight.)

The actual calculations of the abundances depend not only on the solar measurements but also on properties of the elements themselves measured in laboratories on Earth. These latter studies are known as "laboratory astrophysics" and are fundamental to understanding the Sun. (In 2012, the American Astronomical Society added a division of laboratory astrophysics.) The laboratory data are continually refined, most recently providing good values for the element iron and thus an accurate value for the abundance of iron in the Sun.

The Sun, of course, is making elements deep inside its core even as we speak. These elements are not fully mixed throughout the Sun. Thus the elemental abundances we measure for the photosphere are not those for the Sun as a whole, though the general trends are the same.

Table 3.1 shows the photospheric abundances of the most common elements. Note that hydrogen and helium together make up more than 99 per cent of the atoms in the Sun: everything else is a trace element. Indeed, astronomers refer to everything heavier than helium as "metals," simplifying some calculations so that only terms for hydrogen, helium, and metals are used. Note that this use of the word "metal" is different from the electrical-conducting and heat-conducting definition in normal usage. Helium cannot be seen in the solar photosphere, so the abundance in Table 3.1 for helium comes from measurements of the outflow of gas from the Sun known as the solar wind, in which we can detect helium directly. Similarly, the neon measurements are for other astronomical sources.

Table 3.1 The most abundant elements in the solar photosphere.

For each: 1,000,000	atoms of hydrogen
There are: 98,000	atoms of helium
850	atoms of oxygen
400	atoms of carbon
120	atoms of neon
100	atoms of nitrogen
47	atoms of iron
38	atoms of magnesium
35	atoms of silicon
16	atoms of sulfur
4	atoms of argon
3	atoms of aluminum
2	atoms of calcium
2	atoms of sodium
2	atoms of nickel

Motions on the Sun's surface

Colors are so vaguely defined that you can't really tell when they change slightly. But Fraunhofer lines are so clearly and sharply defined that you can measure small changes in their wavelengths. These changes are caused almost entirely by the motion of the gas that is emitting the lines.

The change in wavelength caused by motion is known as the Doppler effect, after the 19th-century scientist Christian Doppler. It is the same effect you hear when a train goes by you, and the pitch of its whistle seems to go down. When the train is approaching you, the pulsations of air coming from the whistle pile up more frequently, and the wavelengths are scrunched up, since subsequent wave peaks are emitted when the train is closer to you than the earlier wave peaks: the frequency is higher and the wavelength is shorter. When the train is receding from you, its wavelengths are stretched, and the frequency goes down, for the opposite reason. So as the train passes, you hear the frequency go from higher to lower. That change in frequency corresponds to a change in wavelength, which we commonly call a Doppler shift.

The corresponding measurement for light is the Doppler shift that moves Fraunhofer lines to slightly longer or shorter wavelengths. Since blue is at the short end of the visible spectrum, changes toward shorter wavelengths – corresponding to motions toward you – are known as blueshifts. Since red is at the long end of the visible spectrum, changes toward longer wavelengths – motion away from you – are known as redshifts. (Even when things are shifted through and past the red to longer wavelengths, they are still known as redshifts because of the term's historical roots.)

Remember that the Sun is so bright that its spectrum can be spread out to a tremendous degree while still leaving enough brightness to be recorded in our measuring instruments. When a solar Fraunhofer line is looked at in great detail and at high magnification, you can see that some bits of it – corresponding to different nearby locations on the surface of the Sun – are shifted to longer wavelengths and other to shorter wavelengths. Thus we speak of "wiggly lines" – a technical term in this context (Fig. 3.7). These wiggles are really just Doppler shifts from bits of gas on the Sun coming toward us (providing blueshifts) or going away from us (providing redshifts). We measure velocities of about 2 km/s for the wiggly lines. This is a low velocity for astronomy, though it is still 7200 km/hr, about 10 times the speed of a passenger jet. These wiggly lines are a clue that there is some type of organized motion in the photosphere, with masses of hot material moving both toward and away from us.

Fine structure of the solar photosphere

If you look in detail at the solar photosphere on a day when the air above is steady (we say that "the seeing is good," using the term "seeing" in a technical sense to indicate the steadiness of

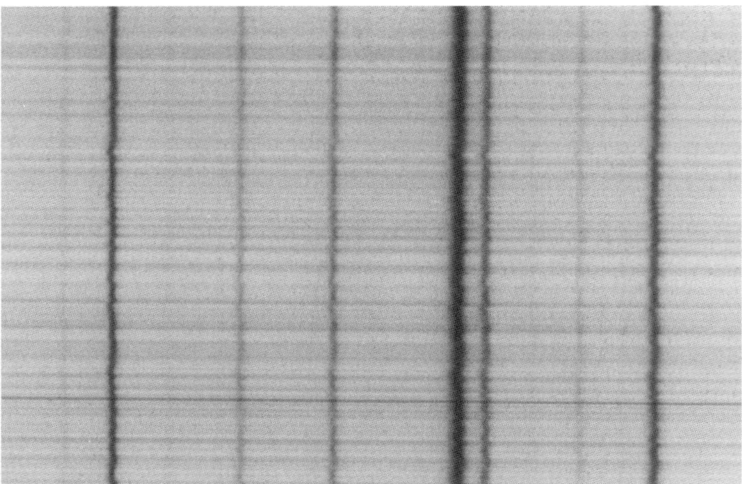

FIGURE 3.7. A spectrum showing left and right displacements along the spectrometer slit, due to motions of photospheric granules on the solar surface.

the image), you see a roughening of the surface, known as "granulation." These granules are at the limit of what you can see in size, given the turbulence of the air above you, except at times of unusually good seeing. This turbulence normally limits you to about 1 second of arc (1/60th of a minute of arc and therefore 1/3,600th of a degree of arc), which corresponds to about 700 km (400 miles) on the Sun, about the distance from New York to Washington. Imagine the entire Eastern seaboard of the United States blurred out to a single unresolved blob and you will get an idea of the minimum size of what we can routinely see on the Sun. There are about 5,000,000 of these granules on the Sun at each moment.

The granules also turn out to be typically about 1.5 second of arc across, which means that seeing them in detail is at the limit of our capability; many of the granules are smaller. One technique used to examine them is to take a lot of short exposures very rapidly, and then selecting the few for which the seeing

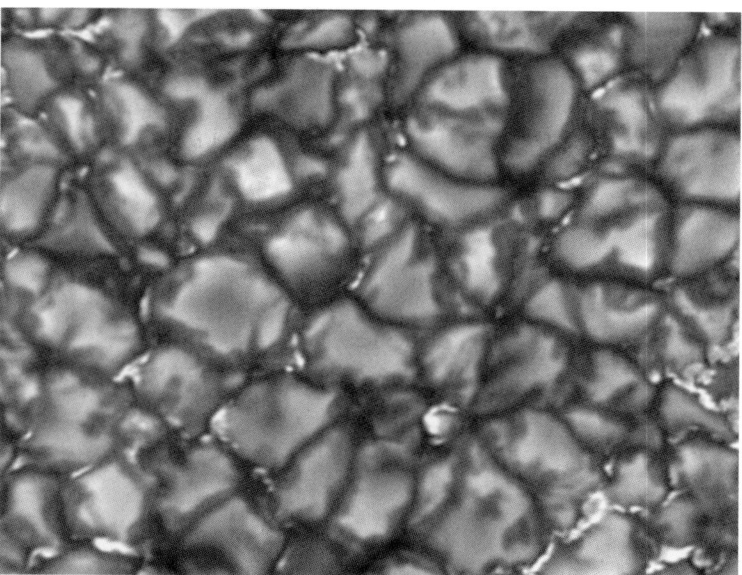

FIGURE 3.8. A high resolution close-up of the solar photosphere taken in visible light, showing the solar granulation and the small magnetic elements called filigree.

was exceptionally good. Such images show bright granules with dark lanes between them (Fig. 3.8). The dark lanes may be only 1/4 arc second across. Sometimes the highest resolution images show bright points along the dark lanes. These bright points, known as "filigree," seem to be 100 degrees Celsius or so hotter than the surrounding surface, and may be linked with the magnetic field, showing how finely broken up the magnetic field is as it pokes through the solar surface.

The granules turn out to be blobs of gas that have just risen from lower levels of the Sun, from below the photosphere. These blobs carry energy upward, just as bubbles of water in a boiling pot carry energy upward. Thus granules are a boiling phenomenon on the Sun, with their bases perhaps only 100 miles down. After the granules reach the surface, they cool and sink back down, only to be replaced by new granules after about 15 minutes. We shall see later on how granules seem to oscillate in brightness with a period of 5 minutes. That oscillation was once thought to be a minor effect, but we will see that it and other, related, oscillations are the key to understanding the Sun's interior.

To see the granules steadily, we must be in space, where atmospheric effects don't cause seeing problems. A mission carried aloft aboard a Space Shuttle, for example, provided an excellent series of images of granules and how they appear and then move slightly across the photospheric surface to merge with their neighbors. These images were used to help understand how convective motions evolve in the complex conditions of the Sun, where the temperature and density of the atmosphere change very rapidly with depth. One of the memorable features of these data was the observation that some of the granules were seen on the Space-Shuttle movies to seem to explode, with debris going out to the sides.

It is also possible to obtain granulation movies of very high quality during rare periods of exceptionally good atmospheric seeing. A sequence of such images (Fig. 3.9), taken at 3.5-minute intervals at the Swedish 1-m Solar Telescope (which, in 2013, became part of Stockholm University instead of the Royal Swedish Academy of Sciences), show small "bright points" in the lanes between granules. These points correspond to strong concentrations of magnetic fields, and the sequence shows that they are buffeted and shredded by the convective motions. This type of rearrangement of the magnetic fields may be connected to the heating of the overlying corona.

An image of the Sun, even one merely projected onto a piece of paper, also reveals slightly bright regions, irregular in shape, which are best seen in the outer parts of the disk, superimposed on the limb-darkened region just inside the edge of the disk. These regions are known as "faculae." Each facula corresponds

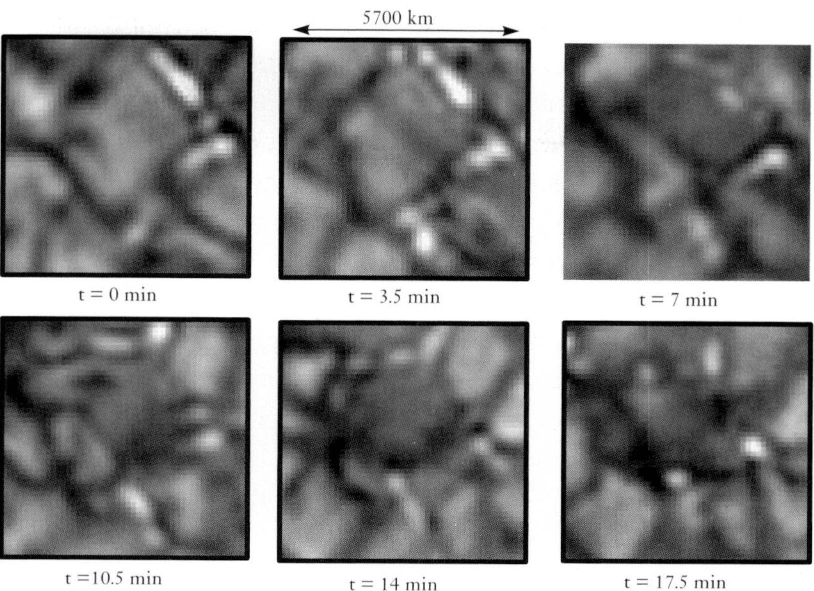

FIGURE 3.9. A granulation photo sequence, showing the buffeting endured by small magnetic elements at the solar surface.

to a region on the Sun where the magnetic field is slightly higher than surrounding regions. Many of the terms like facula come from the early years of the 20th century when George Ellery Hale and his colleagues were examining the Sun in detail for the first time.

We will not detail all the individual names for phenomena on the Sun, but faculae are historically important, for a reason unrelated to solar activity per se. In the 1960s, careful measurements seemed to indicate that the Sun was not quite round, with its equator bulging about 35 unaccounted-for kilometers. A team of scientists incorrectly deduced that the small extra brightness they found near the Sun's equator implied that the equator was bulging by a few kilometers. They inferred from those extra kilometers that the inside of the Sun was rotating much faster than the outer layers that we can see. This extra rotation would, in turn, have an effect on the orbit of the planet Mercury. But we don't see such an effect on Mercury! Thus the scientists concluded that Einstein's general theory of relativity, which otherwise explains Mercury's orbit nicely, had to be modified. They then had their own alternative theory waiting. It turned out, though, that their measurements that the Sun was not quite round were really measurements that the Sun's edge was slightly brighter near the equator (causing it to look "fatter") than near the poles, and that extra brightness was caused by the faculae. So the study of solar faculae turned out to have had major consequences for the fundamental theoretical framework that we use to understand our Universe.

Supergranulation

The granules don't just go up and down by themselves. They have a pattern of flow across the surface of the Sun – that is, we find that the granules form within a larger-scale pattern known as supergranulation, each cell of which is about 20,000 miles (30,000 km) across – about 2½ Earth diameters in size. These supergranules last a day or two, much longer than the 15-minute lifetimes of the individual granules. They are often spoken of as polygons, since their shapes are not just round but neither are they very irregular.

The supergranules' horizontal motion is about 1000 feet/s (0.4 km/s). When the granules move outward from the center of a supergranule toward its edges, they carry along bits of the Sun's magnetic field and this magnetic field piles up at the edges of the supergranules. Supergranules thus show up in magnetic maps of the Sun as an irregular pattern, like a network, of concentrated magnetic field.

The supergranules also show up in velocity maps of the Sun, (Fig. 3.10), which is how they were first discovered. There is a slight upwelling at the centers of supergranules, then the

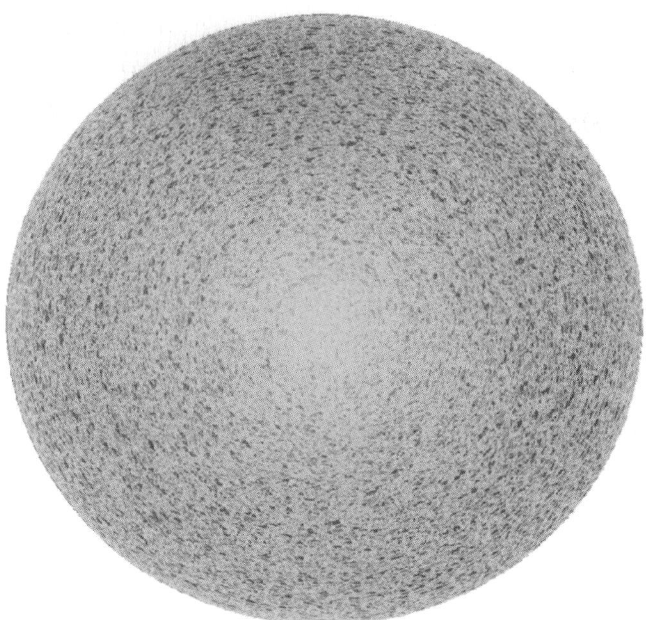

FIGURE 3.10. A velocity map of the solar surface, showing the large-scale flow pattern known as supergranulation.

horizontal flow, and a downdraft at the edges, perhaps only 1/4 the velocity of the upwelling. Doppler images show the horizontal flows, bright for motion toward the Earth and dark in the away direction. At the center of the image, when we are looking straight down at motions that are mainly perpendicular to the viewing direction, there is nearly no Doppler shift and the supergranular pattern is hard to see.

The Global Oscillation Network Group (GONG) is a ground-based set of telescopes that are positioned so that, at one station or another, they are continuously monitoring the Sun. Their data for the solar surface can be broken down into several different components, some of which show the velocities and thus the supergranulation (Plate II). These components are, respectively: I, an image of the white light intensity; D, the intensity variation after the overall brightness variations have been subtracted, as measured by the darkness of a chosen absorption line; V, the velocity field, measured by the Doppler shift. This shows mainly the solar rotation, which is toward us on the left and away from us on the right, but close examination also shows the supergranulation pattern; and B, the magnetic field strength, measured via the Zeeman effect.

One of the instruments on board the Solar and Heliospheric Spacecraft (SOHO, a joint ESA/NASA project) found that the outflow of hot material from the Sun known as the solar wind seems to be emitted through the lanes between convection cells, essentially through cracks in the solar surface.

Quiet Sun vs. active Sun

The phenomena we have just described – including the photosphere, the granules, and the faculae – are on the Sun every day. We speak of the underlying "quiet Sun" that has these phenomena.

Sunspots, on the other hand, wax and wane with an 11-year period. They and their surrounding active regions, discussed in Chapter 2, are known as the "active Sun." So solar astronomers often talk of a basic quiet Sun with an overlay of active Sun.

The quiet Sun can be well studied in the visible part of the spectrum. But the photosphere is too cool to give off x-rays, and does not give off radio waves, so does not appear in other parts of the spectrum than the visible. On the other hand, some of the other parts of the Sun are best studied in these other parts of the spectrum, as we shall see.

The next sunspot cycle

Though it is notoriously difficult to predict the solar-activity cycle into the future, discussions at the 2011 meeting of the

Solar Physics Division of the American Astronomical Society gave some interesting clues. Several lines of evidence seemed to indicate that the next sunspot cycle will be relatively weak or even absent. (See http://www.skyandtelescope.com/community/skyblog/newsblog/123844859.html)

The main point is that even though the cycle that is peaking in 2013 or 2014 is still going on, precursors of the next cycle, Cycle 25 in the current labelling, should have occurred.

One line of evidence had to do with helioseismology, which shows bands of circulating material below the solar surface, a sort of jet stream about 7000 km below the solar surface. The high-latitude band typical of the new cycle hadn't appeared yet, in measurements by Frank Hill of the National Solar Observatory, even though such an earlier appearance of a high-latitude band had been expected from its appearance at the previous cycle.

Further, William Livingston and Matt Penn of the National Solar Observatory have been measuring the strength of the magnetic field in the middle of sunspots, their umbrae. They find that the strength of the magnetic field has been declining over the last decade. When the magnetic field in umbrae falls below about 1500 gauss, which would happen in about 2022 at the current rate, they expect that no sunspots would form! As with so many things about the sun, we will have to wait and see what actually happens. Whether the cycle is normal, weak, or absent, or even strong, we will learn something interesting about the sun.

THE CHROMOSPHERE

At a solar eclipse, just after the Moon entirely covers the solar disk, a reddish rim becomes visible. It is one thousand times fainter than the photosphere but still one thousand times brighter than the corona that will soon appear. This reddish rim is a shell of gas around the photosphere. From the Greek words "chromos," meaning "color," and sphere, it is called the chromosphere. The chromosphere is only about 10,000 km thick, less than 1 per cent of the 1,500,000 km diameter of the Sun.

The chromosphere at the Sun's limb

At the beginning of the total phase of an eclipse, when the photospheric light fades enough, the chromosphere becomes visible even to the naked eye. In the previous section, we described how light from a hot source going straight ahead into a cooler gas can be absorbed by certain atoms, giving rise to the Fraunhofer absorption lines. The lines are dark because the cool gas absorbs the light trying to pass through, and then reemits that energy in all directions. So when you look down from above, as we do when we look at the photosphere, you see the missing energy as dark lines. But when you look from the side, as we do when we see the chromosphere at the edge of the Sun, we see that energy emitted without the bright photospheric background. Thus we see a set of bright wavelengths coming from the chromospheric gas.

These wavelengths include all the Balmer lines of hydrogen. The brightest of them is H-alpha in the red, so the chromosphere appears reddish. But the bluer wavelengths of Balmer lines are also mixed in, as are the violet wavelengths of the calcium H and K lines and several other emission lines.

The chromosphere is covered within a few seconds by the advancing edge of the Moon. After the seconds or minutes of totality, the other edge of the Moon uncovers the chromosphere at the other limb of the Sun. When you see that reddish edge just over the Moon's limb, you know that the total phase of the eclipse is about to end. We will say more about eclipses in Chapter 5.

The flash spectrum

The laws of radiation from hot gases are such that the bright chromospheric emission lines are at exactly the same wavelengths as the Fraunhofer absorption lines seen in the spectrum of the photosphere. After all, they are caused by the same atoms, which both take up and give off energy at the well-defined wavelengths.

You can see this effect clearly by taking spectra as an eclipse begins. This technique was often used in the first half of the 20th century but is rarely used any more. Donald Menzel became famous (among astronomers) in 1931 for his analysis of old eclipse observations of the spectrum of the chromosphere (Fig. 3.11). and Menzel went on to make several sets of spectra of the chromosphere himself. One of the techniques was to move the film from top to bottom, given that the spectrum was spread out from side to side. In that way, you got a continuous photograph that showed how the spectrum changed over time. In particular, you can see that the brightest chromospheric lines turn into emission before the fainter lines do so. One of Menzel's most famous methods was the "jumping film" image, in which the film was moved very quickly between one exposure and the next, without a shutter.

As the photosphere wanes, its spectrum remains that of absorption lines, with only a couple of exceptions, such as for the strong H and K lines and for H-alpha. Then, for a few seconds, the absorption lines all across the spectrum turn into emission lines, as the H and K lines and H-alpha had a few seconds earlier. This emission spectrum seems to flash into view and then disappear again, and has the name of the "flash spectrum." The flash spectrum appears because the bright photospheric Fraunhofer spectrum that overwhelmed it is covered by the Moon, and it disappears because the chromosphere itself is then covered by the Moon. A similar sequence appears in reverse at the end of totality.

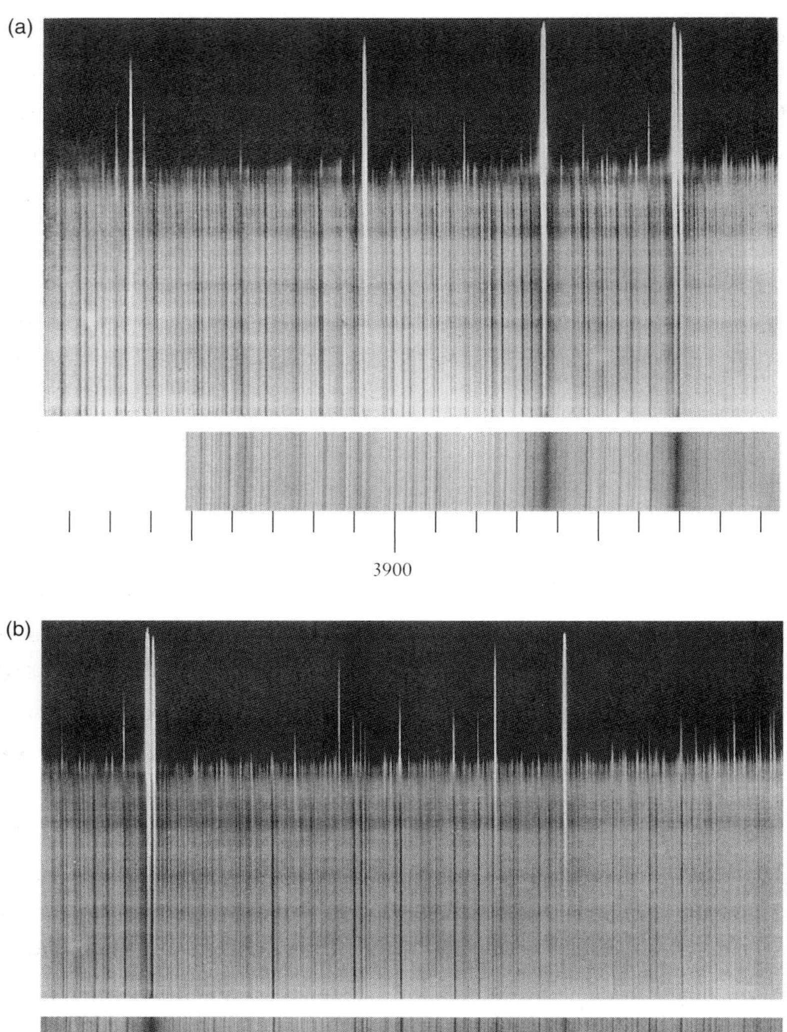

FIGURE 3.11. Moving film image made by W. W. Campbell at the 1905 eclipse in Spain: the film was moved vertically during the observation, the spectrum from a narrow strip of sun being exposed at any time, so the vertical axis shows how the spectrum changed over time at the onset of totality.

The discovery of helium

People began taking spectra at eclipses in the 1860s. At the eclipse of 1868, P. J. C. Janssen of France was watching the flash spectrum with a spectroscope – where "scope" instead of "graph" simply means that he was looking through it with his eye. During the flash spectrum, he noticed several bright emission lines. One of them was yellow, so at first he thought that it was the D line, one of Fraunhofer's original lines then already known to actually be a pair of lines close together. This pair of lines was known as D_1 and D_2. It resulted from sodium, and you can reproduce them easily by throwing a pinch of salt into a flame on your stove.

Janssen realized that this and some other lines of the flash spectrum were so bright that he might even be able to see them without an eclipse. He thus tried to do so the next day and succeeded in seeing it shine outside the solar limb. A few months later, in England, Norman Lockyer also tried and succeeded in seeing chromospheric spectral lines outside the solar limb with a spectroscope even without an eclipse. Coincidentally, the letters of Janssen and Lockyer announcing their work arrived at the French Academy on the same day, forever linking their names (and their faces on a joint medal) with the discovery of the gaseous nature of the chromosphere, rather than merely the discovery of what was later named helium. Other scientists, specifically Norman Pogson in India, might also be credited with the incipient discovery of helium.

With the ability to study the flash spectrum outside of eclipse, Janssen and Lockyer realized that the yellow line they saw in the solar spectrum was slightly displaced from the D_1 and D_2 lines. Following correspondence with Lockyer, the Italian astronomer Father Pietro Angelo Secchi called it D_3. Though the D_1 and D_2 lines were known to come from sodium, the D_3 line didn't seem to come from any known element. Later, a chemist colleague of Lockyer gave it the name "helium," from the Greek word "helios," the Sun. It wasn't until 1895 that helium was extracted chemically from other terrestrial gases. We now know, of course, that helium is one of the fundamental elements, the second least massive nucleus after hydrogen. This discovery was perhaps the most significant ever to come from a solar eclipse.

Technically, the D_3 line is at 587.6 nm, while the D_1 and D_2 lines are at 589.6 and 589.0 nm. They are easy to tell apart when they both appear on a spectrum of sufficient quality, although the D_3 line is still hard to study. Helium needs the gas to be at a high temperature for it to produce this line, and the photosphere just isn't hot enough, so the helium D_3 line shows only weakly in the photospheric spectrum. It is therefore nearly impossible to make an image of the solar disk in this helium D_3 line. But another helium line exists at 1083 nm, in the infrared and thus beyond the spectral region in which film is sensitive. Now helium 1083 nm is routinely mapped with telescopes on Earth using electronic detectors sensitive to the infrared, and you can call up such an image on the World Wide Web on most days (at http://umbra.nascom.nasa.gov/images/latest.html; also at http://mlso.hao.ucar.edu/), if bad weather does not prevent the observing (Plate III).

The chromosphere on the disk

We mentioned earlier that we see into the Sun until the opacity is too great for us to see any further. If we look in white light – all the Sun's light taken together – or in any continuous set of wavelengths, the limit of our vision is some height in the solar photosphere. Everything between it and us is pretty transparent.

But if we look at the wavelengths near the center of the strongest spectral lines typical of the Sun – mainly the calcium H and K lines and H-alpha – the Sun looks very different. The opacity at these wavelengths is very high because the light is strongly absorbed. Thus we don't see in as far. In fact, our line-of-sight stops somewhere in the chromosphere, and we can obtain an image of the chromosphere by putting a narrow passband filter in the optical path of the telescope.

It is not as easy to make such images of the chromosphere as it is to make images of the photosphere. We need to limit our vision to a very narrow band of wavelengths at and around the central wavelength of the strong emission lines. Normally, we want only 0.1 nm or so around the desired wavelength.

Traditionally, for the last hundred years, the device used to make such observations was the spectroheliograph – which literally translated means a spectrum-Sun-writer. In a spectroheliograph, the image of the Sun falls on a slit that extends across the Sun from top to bottom. Light coming through this slit is spread out in wavelength, usually so that color goes from side to side and position along the slit goes from top to bottom. In order to produce an image of the Sun in the desired narrow wavelength range the trick, then, is to have a second slit that passes only the solar light of whatever wavelength you choose. By making that second slit sufficiently narrow, you can allow as narrow a set of solar wavelengths to pass as you choose – as long as you lengthen your exposure time enough so that your film or other detector records enough light. This produces a spectrally pure image of the Sun, but it is only a long, thin piece, basically as wide as the slit and as high.

The further key trick is then to move the pair of spectroheliograph slits from side to side. The slits both move together, so that as the first slit scans the whole Sun from side to side, the second slit continues to pass the desired wavelength for each part of the Sun in turn. As a result of this scanning, an image of the Sun is mapped out in whatever wavelength was chosen. Often that wavelength is the H-alpha line or the calcium H or K line (Figs. 3.12a and 3.12b).

These spectroheliograms show the solar chromosphere. The stronger the absorption line, the higher the opacity and the

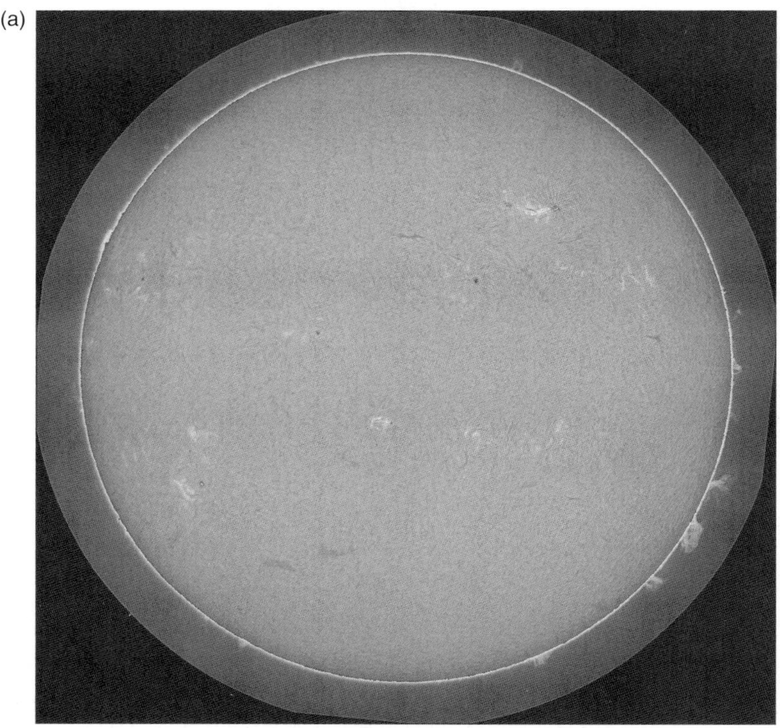

FIGURE 3.12. a) H-alpha and b) Ca K-line images of the Sun, 3 January, 2013.

higher up you see in the chromosphere. The H and K lines are the strongest, and thus show us the highest level of the chromosphere. H-alpha is not quite as strong, and so shows us a high but not quite as high level of the chromosphere. Thus by looking at various spectral lines, we can make a 3-dimensional map of the chromosphere.

To see the chromosphere, you must isolate the emission line. Thus you can accept only a very narrow band of wavelengths. The problem with the method described above is that it is very slow, requiring up to an hour in some cases to produce an image of the entire Sun. In the 1930s in France, the astronomer Bernard Lyot worked out a way of making a filter that passed

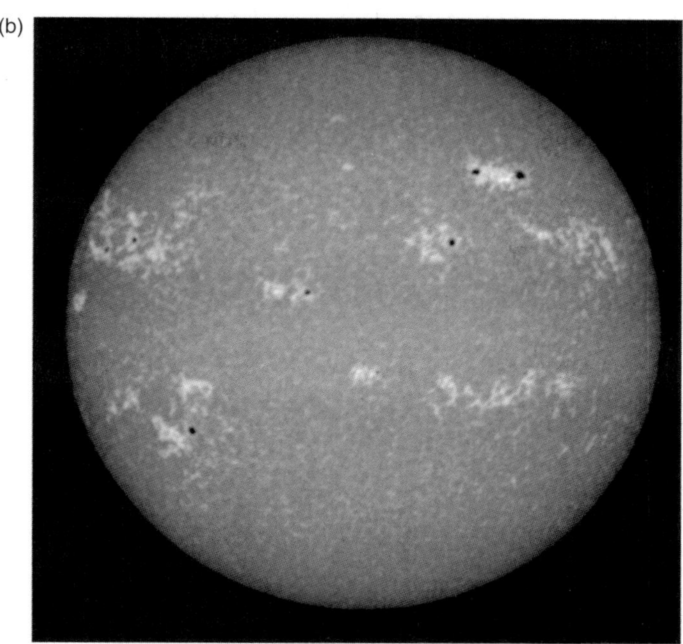

FIGURE 3.12. *(cont.)*

only such a narrow band of wavelengths. These filters used pure crystals of calcite, which have a property called birefringence in which light that enters breaks into parts that are differently polarized, that are displaced from each other, and that travel through the crystal at different speeds. Combining several crystals of calcite of different lengths, Lyot succeeded in making the filters needed to observe narrow wavelength bands. These filters are known as birefringent filters or, often, Lyot filters. They have excellent optical quality and have what is known as square passbands. That is, if you graph their transparency against wavelength, the graph shows a constant level of transparency in the central wavelengths with a constant level of opaqueness outside it. Optically pure calcite is rare and expensive, and is harder and harder to find, so that Lyot-type filters may cost $50,000 today. But they can give excellent solar images in H-alpha and in other wavelengths, without the need for rastering or scanning across the disk. Many Lyot filters were limited to a single wavelength each, though some widely adjustable ones have also been built.

A much less expensive way of making a filter with a narrow passband is to deposit thin layers of reflecting material on a clear glass or quartz piece. The light bouncing back and forth between the deposited layers sets up a pattern of optical interference, and only a narrow passband emerges. These filters are much less expensive than Lyot-type filters, but their passbands are not as square. There is often more leakage of unwanted wavelengths, so the resultant image is not as purely from the desired spectral line. Still, these interference filters are so much less expensive and so readily available that they are in wide use, even for amateur astronomers. Both these filters and Lyot filters are usually temperature controlled, since the wavelength changes as the temperature of the filter changes.

The chromosphere from space

Imaging the Sun in H-alpha or in helium in the visible spectrum shows the chromosphere, as we have discussed. But there are more fundamental lines of hydrogen and helium that are in the ultraviolet, at wavelengths far too short to come through the Earth's atmosphere. Fortunately, special spacecraft outfitted to observe the Sun now observe these more fundamental lines on a daily basis.

The helium spectral line in the ultraviolet at 30.4 nm is one of the EUV lines observed with the Atmospheric Imaging Assembly (AIA) on NASA's Solar Dynamics Observatory (Plate IV). The Lyman-alpha line of hydrogen, a spectral line similar to the red H-alpha line but from a more fundamental energy level of hydrogen and thus much stronger, was observed regularly with the Transition Region and Coronal Explorer spacecraft (TRACE, Plate V), which sent back data from 1998 until 2010.

Appearance of the chromosphere and prominences

The chromosphere as seen in H-alpha spectroheliograms or spectrograms shows a lot of structure that isn't seen in the photosphere. Sunspots themselves don't show well in H-alpha, but the regions around them are relatively bright. These regions are known as plage (pronounced plahzh), from the French word for beach.

Within these plage regions in H-alpha we see narrow, dark filaments. These filaments turn out to trace out the boundaries between areas on the solar surface that are dominated by opposite magnetic polarities. That is, when one sees such a dark filament, it means that there is typically a large patch of outwardly-directed magnetic field on one side of the filament and inwardly-directed magnetic field on the other side. This implies that the magnetic field arches over the filament, and there is indeed good evidence that this is the case.

When the filaments rotate so that they are on the edge of the Sun, they stick up off the edge. We then see them emitting H-alpha, and call them prominences (Fig. 3.13). Prominences can be very stable, and can last for months, though some do erupt in periods of hours. Although they stick up into the hot corona, they are made of cooler gas at chromospheric temperatures of 10,000 kelvins or so. Because they are both cool and stable, they are thus very different from solar flares, which are explosive phenomena at temperatures of millions of kelvins.

The calcium H-line and K-line images look very different from the hydrogen H-alpha images. Let us talk mainly about the K-line images; the H-line images look very similar, though they aren't always as pure when observed in filtergrams because the H-epsilon line of hydrogen has a wavelength very close to that of the H-line of calcium, and so gets partially included in the images.

The K-line images show polygonal regions that are the supergranules. Aligning them with magnetic-field images taken of the

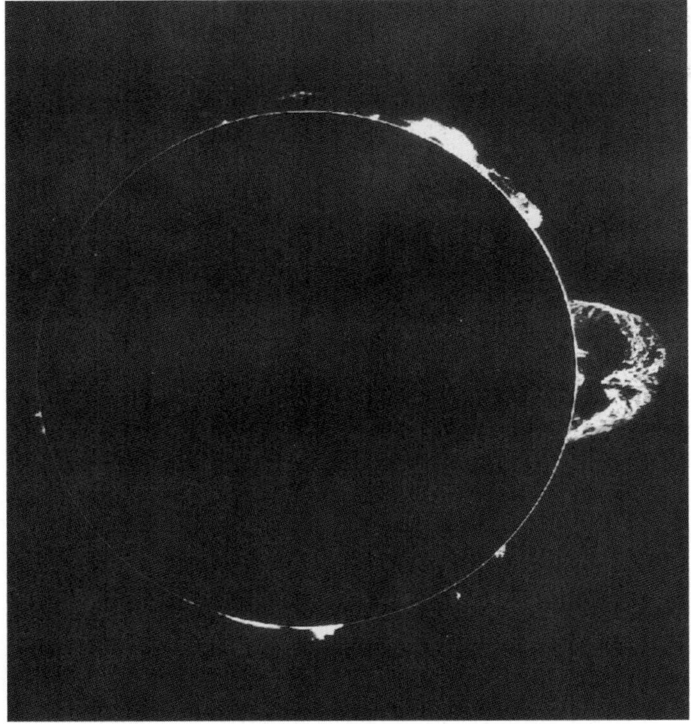

FIGURE 3.13. H-alpha image of a large prominence, 13 March 1970.

photosphere shows that the supergranule boundaries, 30,000 km or so across, match regions of moderately higher magnetic field. These magnetic fields at the supergranule boundaries are noticeably stronger than the average field in the middle of the supergranules, but still substantially lower than the magnetic fields in sunspots. The boundaries are known as the chromospheric network.

Spicules

If you look carefully at the limb of the Sun in H-alpha or some of the other strong spectral lines, you see that it is made up

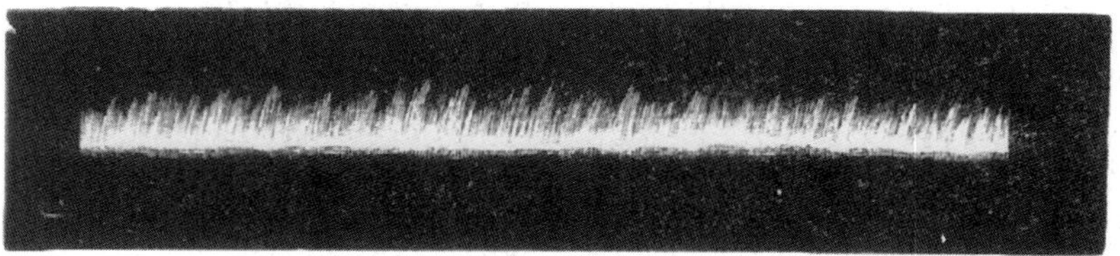

FIGURE 3.14. Drawing of spicules made in 1869 by the Italian Jesuit Father Angelo Secchi.

of little spikes. They have at various times been called "burning prairies" or "leaves of grass," and observers such as Father Secchi in the 19th century invested much effort into drawing them (Fig. 3.14). Since the mid-twentieth century, they have been called "spicules."

These spicules have been studied exhaustively, and can be seen to rise and fall with periods of about 15 minutes. But the individual spicules are only about 1 or 2 arcseconds in diameter, so are barely visible and blur with neighboring spicules, especially because when we view them on the limb each of our lines of sight no doubt goes through several of them. There are hundreds of thousands of spicules on the Sun at each moment, and there is still no general agreement about what causes them.

The spicules consist of columns of chromospheric-temperature material moving upward, and adding all of them together, they carry enough matter to replace the corona, the next higher level, in only a few minutes. Thus they cannot simply carry material upward, and most of them must fall back down, even if we can't see it in every case. Indeed, just watching spicules seem to fall doesn't prove that they are really falling, since they could merely be becoming invisible at their tops as their chromospheric-temperature gas is heated up, and therefore stop emitting in H-alpha. This is a general problem peculiar to narrowband spectral observations: something can seem to be moving in your image either by changing temperature along its length, or by actually moving. Only spectra, on which Doppler shifts can be measured, prove the case for the rising and falling of spicules by showing the velocity of the moving material.

Some objects on the limb, seen especially in helium observations from space, are larger than normal spicules and are known as macrospicules. They have been found to coincide with very small flares seen in x-ray images and this might indicate that the spicules come from yet smaller flares.

The Solar Dynamics Observatory makes steady, high-resolution images of the full solar disk including the solar limb, in which many spicules are visible because the cool, dense spicules absorb the EUV light that SDO is imaging. Spicules are therefore detected in the EUV as dark features (Fig. 3.15) sticking up into the bright corona.

Spicules are important since the chromosphere seems to be made entirely out of them. Though the chromosphere may appear like a layer of gas, and has often been described as a layer and treated as such in theoretical analyses, it is really a

FIGURE 3.15. SDO/AIA coronal image taken January 3, 2013, showing spicules generally as dark absorbing features against a bright EUV background at the limb.

set of vertical spicules each about 1,000 km across and 10,000 km high. If you blur out the spicules and try to calculate temperatures, densities, and other quantities for the chromosphere as if it were a smooth layer, you get an answer, but your answer is not correct. Only by analyzing the spicules individually do you actually find out what the chromospheric gas is like.

For a long time, astronomers tried to trace spicules seen in H-alpha down inside the solar limb to see to what features they connect to on the disk of the Sun. The details of the connection are still unclear, but on the disk, not quite at the limb, one can see structures that look like bushes of spicules (Fig. 3.16). These structures, no doubt, appear as spicules when they are seen at the limb. These structures, however, contain both dark and light mottles, and there is not agreement as to which one corresponds to the spicules. The mottles seen on the disk often appear clumped in radial arrangements known as rosettes, perhaps indicating that they are associated with bundles of magnetic field lines sticking upward and spreading out from a small source.

Mapping the 3-dimensional structure of the chromosphere

We described above how the strong opacity in the center of the K-line of ionized calcium makes our line-of-sight end high in the chromosphere. But a detailed look at the solar spectrum shows that the K-line has a complicated structure in wavelength (Fig. 3.17) which can be useful in making observations at different heights in the chromosphere. Notice that from the continuous radiation (which is known as "the continuum,") that exists at both shorter and longer wavelengths on either side of the K-line spectral structure, there is first a drop down in intensity toward the center of the line. But then, very close to the center, the curve reverses and rises back up. And at the very center, it reverses again, resuming its downward trend. This

FIGURE 3.16. Spicule bushes seen in the wing of the Hα line near the limb of the Sun.

structure is called a double reversal. At the very center of the K-line, the intensity is only about 4 per cent of the intensity of the nearby continuum.

All of these reversals mean that the opacity varies for different wavelengths within the K-line. Thus by looking at different wavelengths in the K-line, you see down to different levels in

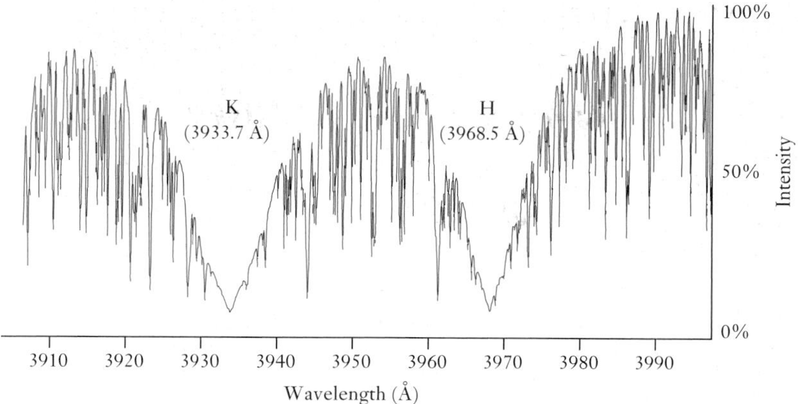

FIGURE 3.17. A detailed look at the solar spectrum in the vicinity of the calcium H- and K-lines.

the solar atmosphere. In the center of the K-line, known as K_3, the opacity is high, and your view terminates at the highest level of the chromosphere. As you go outward to the emission peaks near K_3, which are known as K_2, the opacity is somewhat lower and you see farther into the solar atmosphere, showing a level lower in the chromosphere. As you go through the dips, known as K_1, and farther outward until you reach the continuum, you see still lower and lower levels. By the time you are at the continuum, you are seeing the photosphere. These different levels show in spectroheliograms taken in different parts of the K-line (Fig. 3.18).

Some astronomers spend their time calculating the details of the Sun based on studies of spectral lines, a field of study known as "stellar atmospheres." It is easy to see how as you go inward toward the line center from the continuum, you are looking through gas at higher opacity, and so you are seeing to levels higher than the photosphere from which you start. What is less obvious is that when you get to high levels of the chromosphere, collisions in the gas at some level in the chromosphere cause the

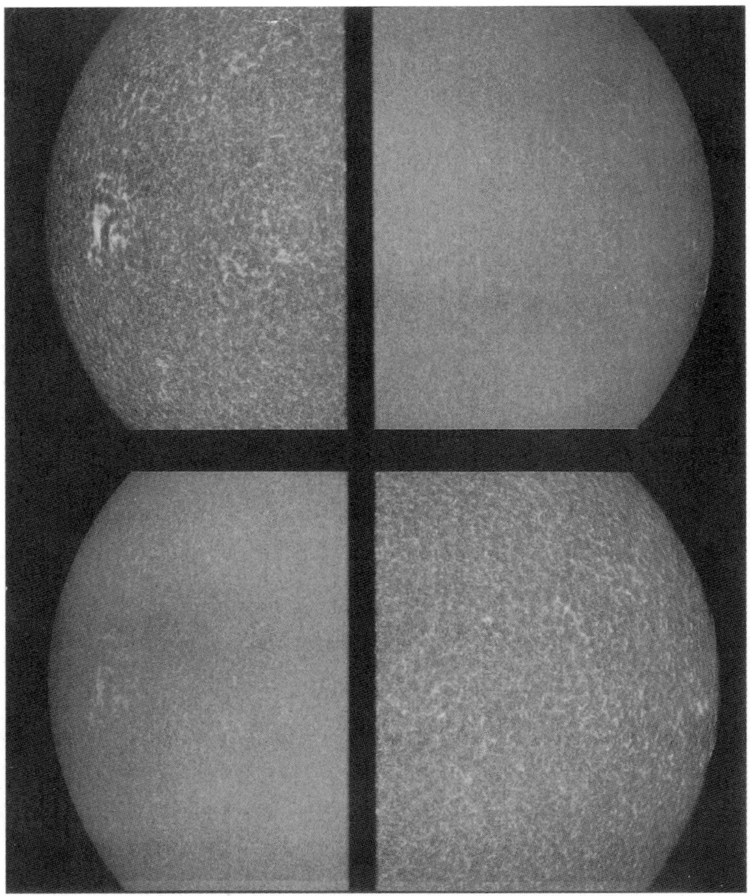

FIGURE 3.18. Different heights in the solar atmosphere are seen as we tune through the K-line. These four spectroheliograms were taken in K_3, K_{1v}, K_{1r} and K_3, respectively.

K_2 peaks to form in the spectrum. At still higher levels, some of the photons of light that are given off can escape readily to the outside. This escape gives you the double reversal in the center of the K line. The calculations involved in understanding how light moving through the solar atmosphere interacts with the gas atoms in that atmosphere – a field of study known as "radiative transfer" – are often made with some of the most powerful supercomputers available. They often assume that the Sun is made of layers; in the calculations, the temperature and density at the top of one layer are forced to match the temperature and density of the bottom of the next higher layer, as must be the case physically and the layers are progressively stacked on top of each other to build up a simulated atmosphere. Although we know that this type of model is wrong for the atmosphere as a whole, it might be a good approach to modelling the properties within each type of vertical structure.

But the radiative-transfer explanation is not the only one. If you take spectra at extremely high spectral and spatial resolution, you can see that the peaks seem to be concentrated at wavelengths slightly to the left or to the right of the K_3 central wavelength. Phenomenologically, one can simply say that these features, which presumably correspond to spicules, are Doppler shifted to the left or to the right because of their motion toward or away from us. The K_3 spectral feature can be caused by a cooler gas with low velocity that surrounds the spicules. Such a detailed look at the fine structure gives an insight that is not available from lower resolution observations. Again, the structure of the chromosphere that you derive from the average blurred image is different from the actual temperature, density, and velocity structure of an individual spicule or even of an average of the values for many spicules. For example, the height of the chromosphere is given, on the basis of average values, as 2000 km above its base, while we clearly see spicules extending to over 10,000 km high.

In the visible spectrum of the Sun, only the H and K lines of ionized calcium are strong enough to show the double reversal. The hydrogen lines, even H-alpha, are not that strong. But as you go to the ultraviolet, there is a pair of ionized magnesium lines at wavelengths just beyond the wavelengths that come through the Earth's atmosphere. When they were first studied well from Orbiting Solar Observatory 8 in the late 1970s, these ionized magnesium lines were given the names h and k, by analogy with H and K of ionized calcium. Historically, of course, it is a very bad notation, and it would have made more sense to extend Fraunhofer's notation to, say, P and Q. But the h and k notation seems to have stuck (Fig. 3.19).

It is particularly important to understand the double-reversal in the structure of the calcium K line because such double reversals are observed in the K-lines of many distant stars. Indeed, the presence of such K-lines is thought to indicate that these distant stars have chromospheres, like the Sun in type but much more powerful. The Sun's chromosphere is so puny that it would not be detected at all if the Sun were as far away as these stars are. The strength of the K-line K_2 reversals varies on the Sun with the solar-activity cycle. Long-term studies of many stars, especially those made at the Mt. Wilson Observatory, have found variations of the K_2 reversals with periods of several years. These stars presumably have sunspot cycles that are causing these reversals.

In the 1950s, Olin Wilson of the Mt. Wilson Observatory and M. K. Vainu Bappu, an Indian scientist who often worked in the

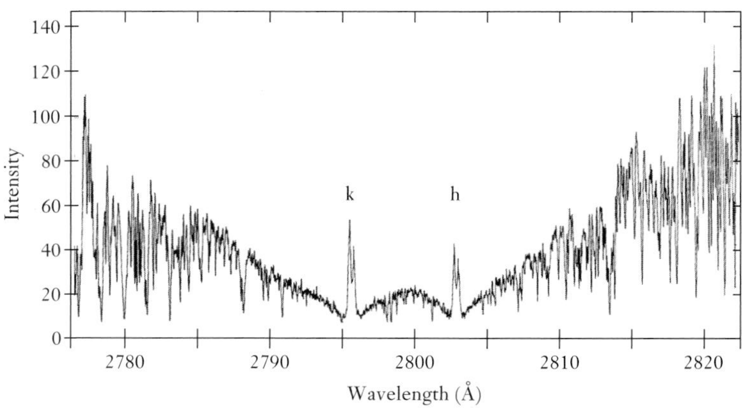

FIGURE 3.19. The h and k lines of magnesium in the ultraviolet.

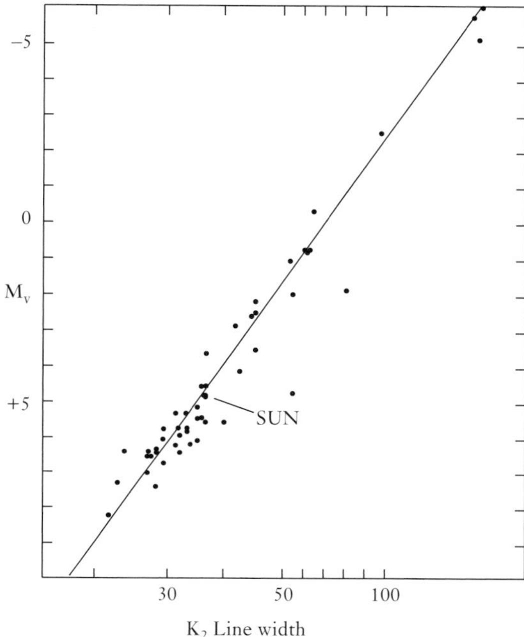

FIGURE 3.20. The strong correlation between the absolute brightness of a star (vertical axis) and the width of the K_2 line in the spectrum (horizontal axis) is known as the Wilson-Bappu effect.

United States, discovered that the width – not the strength – of the central reversal correlated very well with the absolute brightness of the star. Astronomers are always eager to have ways to find out how far away stars are, and this discovery provides a tool for determing distance: by comparing the absolute brightness of the star with its brightness as observed here at Earth, we can calculate how far away it must be. The link between K_2-linewidth and brightness is known as the Wilson-Bappu effect (Fig. 3.20).

Prominences

Prominences are phenomena that vary with the sunspot cycle, and so are considered to be part of the active Sun. The chromosphere as a whole, in contrast, is always there and so is part of the quiet Sun. But prominences are similar in temperature and density to the chromosphere, so they can be thought of as extensions of chromospheric gas higher into space.

Prominences come in many shapes and types, and detailed classification schemes have been worked out for them. A large, long-lasting type is a "hedgerow prominence," and such quiescent prominences extend for perhaps a hundred thousand kilometers along the edge of the Sun (Fig. 3.21). Though we may lose track of it when the rotation of the Sun carries it over the limb and out of sight, it may reappear two weeks later on the Sun's opposite limb and then disappear as it moves onto the face of the disk, and then reappear again on the original limb another two weeks later.

FIGURE 3.21. Hedgerow prominence image.

Other prominences are of the eruptive type. If you watch the Sun through an H-alpha filter for a few hours, you will see these prominences change, rising and falling and sometimes blowing off the Sun into space. Such events may take many minutes or even hours to unfold, and they are not immediately apparent to the eye. To study these events, prominences are sometimes viewed in time-lapse movies, in which one frame is taken every minute and then the movie is played back at 30 frames per second. Thus the motion is speeded up by a factor of 1800, and 30 minutes of time passes by in one second on the screen. Ten seconds, then, shows 300 minutes or five hours of prominence motion, and you see prominences rise, fall, seethe in place, or erupt.

Prominence eruptions are often seen to occur simultaneously with solar flares, but whether or not prominences erupt because of flares is still under debate. Also under debate is the link between eruptive prominences and coronal mass ejections (CMEs), large-scale ejections of hot coronal gas. Such eruptions have in recent years been studied much better than previously because of the wide field of view of the Large Angle Spectrographic Coronagraph (LASCO) on the SOHO satellite and coronagraphs on NASA's STEREO mission twin spacecraft.

Spectra taken of prominences show that they have very complicated motions (Fig. 3.22). In the figure, the vertical axis shows position along the prominence. The white regions show the prominence gas, and whether they are at the left side or right side of the center shows whether their wavelengths are blueshifted (to the left) or redshifted (to the right). The figure shows that the prominence contains gas that is moving rapidly, with both large blueshifts and redshifts, and with a large range of motions at a given vertical position.

The prominences show particularly well in the helium lines mapped by the Skylab Apollo Telescope Mount in the 1970s

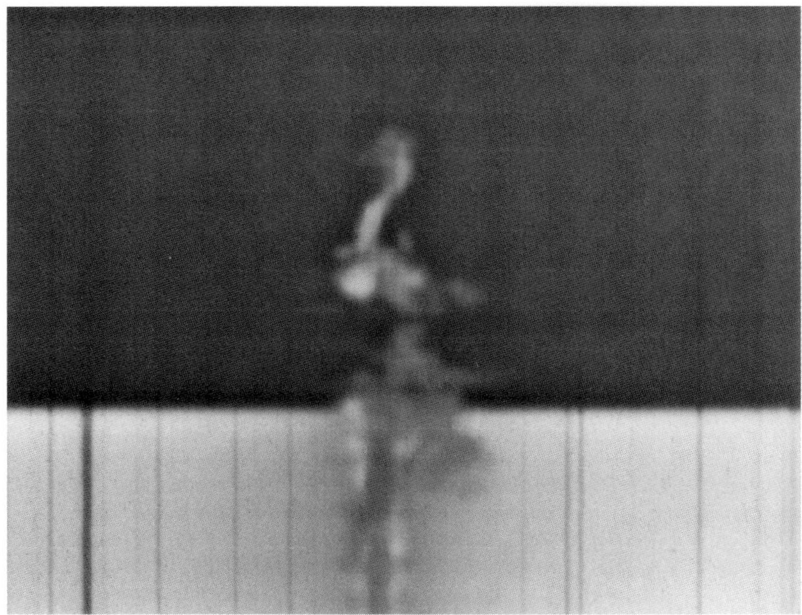

FIGURE 3.22. A spectrum obtained in a prominence shows strong evidence of motions: the image would be a straight vertical bar if there were no Doppler shifts due to motions in the prominence toward and away from the viewer.

and, more recently, from SOHO, Solar Dynamics Observatory and STEREO. In the Skylab images, the ultraviolet wavelength of helium (30.4 nm) overlaps slightly with some coronal images but is basically an isolated emission line. So such images can be taken with spectrographs which form images of the Sun that are positioned alongside each other according to wavelength, sometimes overlapping each other when the images are from emission lines having nearly the same wavelength. For the case of the strong helium line you can, for the most part, just cut out the helium part of the image and ignore the rest (Fig. 3.23).

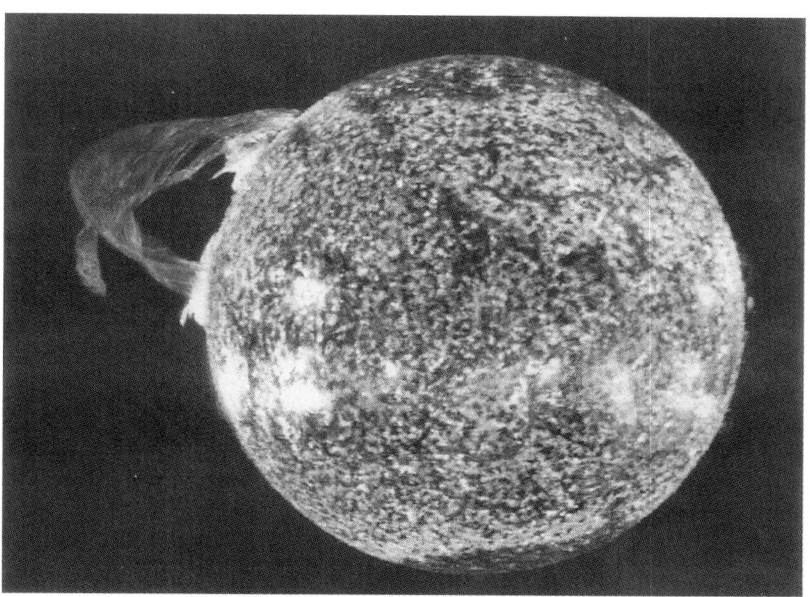

FIGURE 3.23. Skylab image of the Sun in the singly-ionized helium spectral line at 304 Å, showing a prominence eruption.

THE TRANSITION REGION

Between the relatively cool photosphere and the hot corona one finds not only the chromosphere, but also a part of the atmosphere that covers the range from the 10,000 K (roughly 20,000°F) temperature of the chromosphere, to the several million degrees of the corona. This part of the solar atmosphere is usually called the transition region. It does not receive as much attention as the other parts of the atmosphere, and is often thought of merely as an interface between the chromosphere and the corona. However, it is a highly dynamic place, which responds readily to local changes in the heating rate. This makes it ideally suited for studies of the coronal heating mechanisms, and for searches of the source of the outflow of material known as the solar wind. According to atmospheric models, such

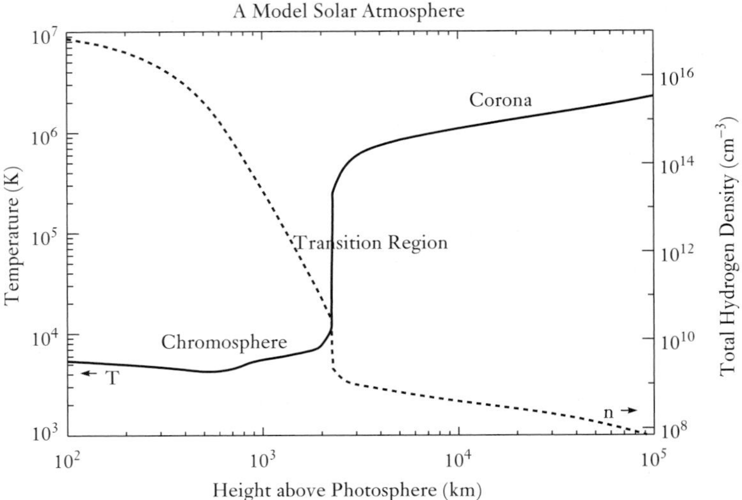

FIGURE 3.24. The temperature of the solar atmosphere as a function of height above the photosphere, for a plane-parallel atmosphere, with no structure in the horizontal direction. The extremely thin region containing the rise from 10,000 to 1,000,000 degrees is called the transition region.

as Figure 3.24, the transition to coronal temperatures happens extremely quickly, over a very short distance.

Figure 3.24 is misleading in at least one respect. This type of display treats the solar atmosphere as if it were a uniform layer, one that changes only as a function of height, but is uniform in the horizontal direction. This is decidedly not the case, as is clear from the photographs we have included in this chapter. The atmosphere is actually highly non-homogeneous, and the temperature is very poorly correlated with height. In particular, we are finding that material at chromospheric and transition region temperatures can be found at great heights above the surface. This material has a complicated fine structure and is often in a highly dynamic state of motion.

A recent article on the transition region listed the following basic questions that remain to be answered:

1. What is the structure of the atmosphere in the transition region, and its relationship to both hotter and cooler structures?
2. Why does the transition region seem to be mostly empty space; that is, why are the structures so small and why do they fill only a small fraction of the volume?
3. How do we explain the distribution of hot and cool material in the transition region, especially the increasing amounts of cool material below 100,000 K, after a decrease is seen between 1,000,000 K and 100,000 K?
4. Is the average model, such as Figure 3.24, an accurate picture of the actual structures in the transition region?
5. What is the heating mechanism that keeps the transition region in its observed state?

These questions remain unanswered at the present time, largely because the transition region is very hard to observe. New and better observations, mainly from space, will be needed for progress to be made. We discuss some of the plans for new observations in Chapter 6.

SUGGESTIONS FOR FURTHER READING

Golub, Leon, and Pasachoff, Jay M., *The Solar Corona*, 2nd ed., (Cambridge University Press, 2010).
Menzel, Donald H., *Our Sun*, 2nd edition (Harvard University Press, 1959).
Noyes, Robert W., *The Sun, Our Star* (Harvard University Press, 1982).
Pasachoff, Jay M., "Resource Letter SP-1 on Solar Physics," *American Journal of Physics*, 78, 890–901, 2010.
Phillips, Kenneth J. H., *Guide to the Sun* (Cambridge University Press, 1992).
Wentzel, Donat, *The Restless Sun* (Smithsonian Institution Press, 1989).

4

What We Don't See

From the evidence brought to light by research in archaeoastronomy, it seems that humans have been constructing instruments to supplement their sensory equipment for many thousands of years. Stone markers, crude sighting devices, and methods for keeping track of monthly and seasonal events were in common use worldwide. The culmination of these naked-eye observations was Tycho Brahe's Uraniborg Observatory – located on an island near Elsinore castle in Denmark – that obtained planetary orbit determinations near the end of the 16th century so accurate that Kepler was finally able to figure out the true shape (elliptical) of the planetary orbits.

But it was Galileo's use of the telescope a few years later that brought home in dramatic fashion the realization that there are strange and wonderful phenomena in the heavens that we cannot see with the naked eye. The phases of Venus, craters on the Moon, moons circling the planet Jupiter, and details of sunspots on the face of the Sun were among the new discoveries revealed by this instrumental extension of the human apparatus.

We have been constructing bigger and better telescopes ever since, continuing right up to the present day. But we have also discovered that restricting our attention to the wavelengths that our eyes can detect is a major limitation. Astronomical objects in general, and the Sun in particular, look markedly different at radio, infrared, visible, ultraviolet, and x-ray wavelengths. By exploring those differences we can begin to understand what it means for the Sun or a star to look so different in x-rays or radio than in visible light.

THE NON-VISIBLE

The light that we see is only a small fraction of the total emitted by astronomical objects. Most wavelengths of light are not able to penetrate our atmosphere to reach us down at ground level, as shown in Figure 4.1. Short-wavelength ultraviolet (UV) and x-ray light are absorbed before reaching ground level, as is infrared (IR) light. It is surely no accident that the light that reaches us through the atmosphere happens to coincide with the light we use for seeing – that is, our eyes adapted to work with the light available (although it is not at all obvious that our atmosphere had to be transparent at the dominant wavelengths emitted by the Sun). The objects that we want to study – the Sun, planets, stars, galaxies, black holes – do not know anything about our atmosphere and much of their energy is emitted at wavelengths that do not survive to ground level. The only way we can detect these wavelengths is to put our equipment above the Earth's atmosphere.

THE SOLAR CORONA

Above the visible surface of the Sun a strange thing happens, something that puzzled astronomers for more than fifty years. Imagine the following: you turn on the burner of your stove,

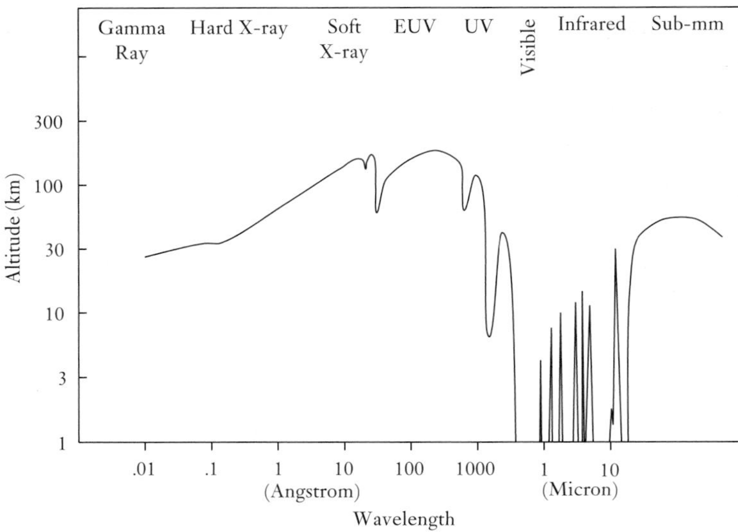

FIGURE 4.1. This figure shows the transmission of the Earth's atmosphere for wavelengths of light from x-ray to radio. The curve shows the altitude at which $\frac{1}{2}$ of the incoming light is absorbed by the atmosphere. Only in a few small "windows" in the visible and near-infrared does the radiation reach ground level.

and then explore the air just above it. As you might expect, just above the surface of the burner the air is hot, but not quite as hot as the burner itself. You probe a fraction of an inch higher and the air has cooled a bit more, although it is still hot. So far, all is well, in that the burner is heating the space around it, but not to a temperature higher than its own. But imagine that you then move another fraction of an inch higher and suddenly the temperature shoots up, becoming ten times as hot as the burner. Something very unusual would need to be happening to produce that effect.

This unlikely situation is what we find in the atmosphere above the visible surface of the Sun. The corona is very tenuous and would qualify as a very good vacuum in any Earth-based laboratory, but it is very hot, reaching temperatures higher than

5,000,000 F only a short distance above the 10,000 F surface. The bad news, from the viewpoint of an observer trying to understand what is happening, is that the extremely high temperature of the corona causes the main part of the radiation that it emits to be not ordinary visible light, but x-rays. As an object heats up, the color of the light it emits moves from the infrared, to the red, then into the violet. In the corona, the temperature is so high that the average energy of a photon of light emitted by the gas is increased so much that the "color" has been shifted far into the blue, into the ultraviolet and extreme ultraviolet. A small amount of visible photospheric light makes the corona visible under some circumstances, but the primary emission from the corona is at short wavelengths beyond the visible spectrum.

In Figure 1.11 we showed a ground-based, visible light image of a total eclipse. An eclipse with the hot corona inserted into the picture, is shown in Plate VI. The inner portion shows emission from the million-degree corona, obtained from a telescope that images the extreme ultraviolet emission of this hot gas directly.

Even if these short wavelengths did reach the ground, we would not see them because our eyes are blind to those colors. So the difficulty is dual, because we need to build instruments that can see colors that humans normally cannot see, then we have to put those instruments up above the atmosphere and somehow record the images and get them back down to the ground. Some of the special difficulties encountered in putting instruments in space are discussed in Chapter 6.

The advantages of looking at the Sun in these invisible wavelengths is clear, even from a single "snapshot" view. Plate VII shows a solar active region viewed in the emission of nine times ionized iron, meaning that the atoms are so hot that they crash into each other forcefully enough to knock the first nine of iron's twenty-six electrons completely out of the atom. Such highly ionized atoms are present when the coronal gas is at a temperature of about 1,000,000 K (kelvins). This type of image cannot be obtained from the ground, and reveals a Sun that is markedly different from the bland white-light photosphere.

The explanation for this complex, hot and dynamic corona still eludes us after nearly seventy-five years of effort. It is almost certainly the case that a central feature of the answer lies in the magnetic fields that are generated inside the Sun and then bubble up to the surface. This is unfortunate for scientists trying to explain the corona in simple terms, because most people do not have any intuitive feeling for the properties of magnetic fields. In addition, the high temperature of the corona means that the coronal gas is ionized – the electrons of its atoms are mostly separated from the nucleii – and most humans have no significant experience of this fourth state of matter known as plasma.

Plasmas are inherently complex because the electrical charges of the particles become important in their interactions, and magnetized plasmas are enormously more complex because the motions of the charged particles are altered by the magnetic field. Moreover, the magnetic field is anchored in the relatively dense solar surface, where the motions of the solar matter are strong enough to affect the field. That field in turn reaches out into the corona, where the coronal matter density is so low that the magnetic field dominates and constrains the gas. The physics of these two regions are quite different, but they are connected to each other by the fact that the field threads from one to the other, allowing the gas to flow along the field from one region to the other. It is not too surprising that we have been making only slow progress in understanding the solar corona.

Solar flares

On the cosmic scale, solar flares are not very big events. A large flare might increase the total energy output of the Sun 0.1% for a few minutes, while other solar-like stars in our galaxy produce flares a thousand times larger. At the extreme, there are believed to be explosions – perhaps due to the interaction of two black holes – that liberate in one second more energy than the Sun will emit in 10 billion years. But on our earthly scale, the solar events are enormous, each with enough energy to power all human activities for centuries.

Even though we are more than 90 million miles away, we feel the effects of solar flares. Radio communications are disrupted – one of the first clear indications came during World War II, when a British researcher worried that the Germans had found a way to jam the newly-invented radar. Power distribution networks have been damaged, satellites have been put out of commission, and the radiation from a flare can be fatal to unprotected astronauts.

But what do we know about the causes of these eruptions on the Sun? Why does the Sun produce these sporadic outbursts? Equally as important, why doesn't it produce more and bigger eruptions?

Unfortunately, there are nearly as many flare theories as there are flare theorists, since many different models are compatible with the laws of physics, and the observations are not yet able to rule out all of the possibilities. But progress is being made, thanks to both improved observations and improved theory, and at least a broad consensus is emerging. Most flare researchers agree that the magnetic fields that erupt through the solar surface into the corona are somehow involved. The reason for this agreement, aside from some observational clues, is that an explanation of flares requires some type of energy storage and a mechanism for rapid release of this stored energy. Magnetic fields can provide both, and we also observe drastic changes in the magnetic field during a flare. Figure 4.2 gives one of the most frequently-invoked magnetic field configurations, which we will call the Standard Model.

What this figure shows is an arch-shaped array of magnetic field lines that are supporting a blob of cool, dense

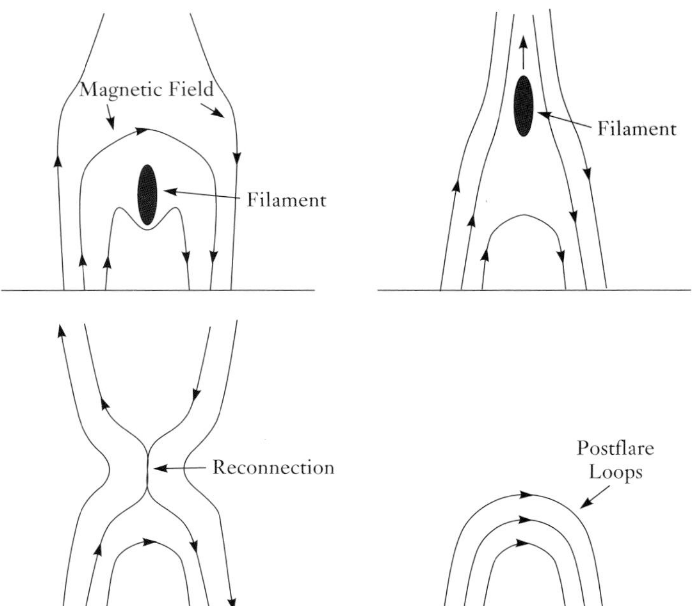

FIGURE 4.2. The Standard Model for solar flares. A mass of cool, dense material being supported by an arcade of magnetic field lines becomes unstable and erupts outward, breaking open the field. Reconnection of the field lines to their closed state releases energy to produce the observed brightenings.

material known as a prominence. (For historical reasons, when a prominence is viewed on the disk of the Sun it is called a filament. When it projects up above the limb and is seen against the dark background of space, it appears bright, and is called "prominence"; when it is viewed face-on against the bright disk of the Sun, it appears dark and is called "filament.") At the top of this arcade the magnetic field is weak and the field lines open up to interplanetary space, forming a shape known as a helmet streamer (because of the similarity to the tops of Prussian military helmets). These structures are seen often in the corona, usually in or near active regions, but there are also larger and smaller replicas of this configuration found throughout the corona.

The situation shown in the top left panel of the figure is dynamic, in that the magnetic fields slowly evolve, and the prominence material flows and moves in all directions, but on the whole it is stable for many hours or days at a time. Occasionally, however, for reasons that are not yet understood (and therefore cause passionate debate among solar researchers) the filament is flung upward and an enormous amount of energy is released, as shown in the next two panels. The magnetic field is broken open, and then connects back underneath the departed filament, and the reconnected "postflare" loops glow brightly at millions of degrees for many minutes or hours. Usually, the region forms itself back into the initial configuration, including the filament, very soon after the flare.

Observations from the TRACE satellite and, more recently, the Solar Dynamics Observatory, show that the Standard Model works well as an explanation for large flares. Plate VIII shows the development of a flare. In the first panel, the flare is just beginning: part of the filament can be seen erupting upward at the left of the frame, and bright ribbons running horizontally across the middle of the frame outline the feet of the arcade of magnetic field lines.

In the subsequent frames, the postflare loops start to form. This particular flare has a second filament eruption occuring in the background – it can be seen moving upward by comparing the top left and bottom left frames – and a second set of postflare loops is visible in the bottom right frame. In this example, the filament and the arcade of magnetic field lines supporting it actually took on a circular, rather than a linear, shape. The postflare loops therefore have a circular configuration, like a bent tunnel. But the principle of magnetic field opening up and then reconnecting underneath is followed in this event.

LOOKING INSIDE THE SUN

Most of the phenomena we observe at the Sun's surface have their origin in the solar interior. In the following we discuss two of the major problems in modern solar interior research: the core of the Sun, where nuclear fusion takes place, and the outer convective envelope, where magnetic fields are generated.

The energy source of the Sun, nuclear fusion, is located deep in the center where temperature and pressure are high enough to ignite nuclear burning. We have very good models of the solar interior, and one result of these calculations is that the nuclear fusion that powers the Sun should also produce a certain number of subatomic particles called neutrinos. But experiments carried out here on Earth do not detect enough of these neutrinos. So there was a problem, usually referred to as "the solar-neutrino problem"; we will shortly see its explanation.

The location of magnetic field generation is difficult to pin down: it may be located anywhere from a shallow surface layer just below the photosphere down to an hypothesized boundary layer between the outer convective envelope of the Sun and a non-convective (radiative) core. Both cases are possible, and finding the right answer involves a delicate interplay between observations and theory that can serve as models for the way in which science operates.

An aside on how science operates: The key breakthrough that made the Scientific Revolution possible at the end of the 16th century was the realization that having a plausible theoretical explanation is not enough, and that a theory must be tested against the actual observed conditions: the famous moment when Kepler realized that the observed position of Mars differed by eight minutes of arc from the position calculated with circular orbits is a prototypical example. Generally, what we need to do is to come up with a theory that not only sounds right, but that actually matches the observations.

Explaining magnetic field generation inside the Sun requires a knowledge of the fluid flow patterns below the solar surface. Only a few years ago it was thought that such knowledge would remain forever out of observational reach, limited strictly to theoretical calculations. However, a key realization – that waves permeate the Sun and can be analyzed by careful study of oscillations at the solar surface – opened up an entire field of study, "helioseismology," that is revealing to us the once-hidden state of the interior.

Neutrinos

One of the great joys of studying science is the occasional rare moment when one is struck by a bit of understanding that completely changes your view of the world. Studying the behavior of neutrinos offers such an opportunity, since they are quite outside of our ordinary experience.

Neutrinos were first postulated on theoretical grounds by Wolfgang Pauli, in order to explain certain nuclear decay processes – that is, ones in which one type of nuclear particle changes into another type – that did not seem to conserve mass and energy. For instance, a neutron just sitting by itself alone in empty space will spontaneously change into a proton, plus an electron. The problem is that the total mass and energy of the observed decay products do not add up to equal the initial amounts of these quantities. More subtly, the way in which the directions and speeds of the decay products (the proton and the electron) are distributed in space seems to violate the mechanical law known as conservation of momentum, unless an additional undetected particle is being produced, if we are to continue believing that momentum is a conserved quantity.

The experiments that Pauli was studying did not show any other particles coming out, but he was not ready to give up on the important laws conserving energy and momentum. So he proposed the "little neutral one" or "neutrino," a particle emitted in these decay events, but which, for some reason, we do not detect. The neutrino remained only a theoretical construct for many years: Pauli proposed the existence of these particles in 1930, and it was the mid-1950s before Clyde Cowan and Fred Reines were able to detect their presence.

The neutrino does in fact exist, and the reason that it is so hard to see is that it does not interact with matter via the electromagnetic force. This plain statement is the one that is at first hard to comprehend: in order for matter to be detectable, that is, in order for it to have visible effects, it must be a "carrier" of a force that we can detect. Put another way, one bit of matter cannot interact with another unless it has within itself the means for interacting. This platitudinous-sounding statement has real implications, because it means that if the particle does not carry the necessary type of force, then it does not interact via that force.

Astoundingly, it seems that there actually are types of matter that do not interact (or hardly interact, to be more precise) with our sensory or observing equipment. This means that they do not affect us or the world around us as they fly by, and this in turn means that we have no way of knowing that they are there. This is a serious problem for science, because it does no good to hypothesize unless there is some way to test the hypothesis; proposing the existence of the neutrino does no good unless there are ways to check whether neutrinos actually exist.

Fortunately, neutrinos do interact very slightly with matter, via a force called the weak nuclear force. This interaction is astoundingly weak compared to the electromagnetic force that accounts for almost everything we see around us. For example, there are roughly 100 trillion neutrinos per second from the Sun passing through your body, but the probability of just one of these actually interacting each *year* with one of your atoms is less than 0.0001% – which is why you don't notice them.

Neutrino detection experiments

Given how hard it is to detect neutrinos – stopping a typical solar neutrino requires about one light-year (5,000,000,000,000 miles) of lead – it would seem to be a practical impossibility to build an experiment to detect them. But one should never discount the cleverness of experimenters, and ways have indeed been found to measure at least a few of the solar neutrinos that are flying past us from the Sun. The longest-running experiment was that of Ray Davis, placed deep in the Homestake mine, and usually referred to as the Homestake experiment. In this setup 100,000 gallons of carbon tetrachloride (ie, cleaning fluid) are used, to take advantage of a reaction in which a neutrino gets absorbed in a chlorine nucleus, converting it to an atom of argon. In chemical-style symbols the reaction is $\nu + {}^{37}Cl \rightarrow e + {}^{37}Ar$. That is, a neutrino hits the chlorine and converts it to argon, with the emission of an electron, which thereby conserves electric charge (and a few other quantum numbers).

The most astounding part of this experiment is that Davis found a way, via chemical analysis, to detect that one atom of argon in this enormous vat of material. Every few months the fluid is sent through the detection equipment and the argon atoms are counted, so that all of the neutrino events during those few months are summed up as a single total number. This is not, therefore, a "real-time" detector, since it might be months after an event before one knows that it occurred. But it is nonetheless an impressive technical achievement.

The main problem with this experiment is that the neutrinos must have a high enough energy to produce the chlorine-to-argon reaction. It requires at least 0.8 million electron volts (MeV), whereas the neutrinos produced by the proton-proton chain – the dominant reaction in the Sun (see Appendix II) – have a maximum energy of 0.4 MeV. The Homestake experiment is therefore sensitive only to the much rarer higher energy neutrinos produced from some of the other less common fusion reactions in the Sun.

The same is true of another experiment, the Kamiokande and its bigger brother, the Super-K, which rely on collisions between incoming neutrinos and electrons in a large batch of ultra-pure water. The interaction is detected by placing an enormous number of light-sensitive photomultiplier tubes around the interaction region (Fig. 4.3). The advantage of this experiment is that it detects the events as they happen, and it can also provide information about the direction from which the detected neutrino arrived. An experiment named Borexino, located in a chamber adjacent to the tunnel passing through the Gran Sasso mountains in Italy, used the scattering of incoming neutrinos from the electrons in a large vat of ultra-pure liquid scintillator surrounded by thousands of photomultiplier tubes. The experiment is far more sensitive than previous ones and provides information on the neutrino energy as well as real-time detection of the neutrinos. Borexino, in 2011, reported direct detection of neutrino oscillations, taking advantage of its sensitivity at lower energies than other neutrino detectors.

Experiments set up in Russia and Italy, known as SAGE and GALLEX, use a neutrino reaction in the element gallium, converting it to germanium. The process is similar to the chlorine reaction, but it does not need nearly as high a neutrino energy. The reaction requires a minimum neutrino energy of 0.2 MeV, so that it is sensitive to the dominant proton-proton neutrinos and the expected neutrino detection rate is some 20 times higher than that of the other experiments. Despite the enormous difficulty and expense of amassing tons of gallium and analyzing it for the occasional germanium atom, these experiments have also run successfully.

The Sudbury Neutrino Observatory (SNO), deep in a mine in Sudbury, Ontario, contains a huge container of heavy water, that is, water in which deuterium is substituted for ordinary

FIGURE 4.3. The Super-Kamiokande neutrino experiment during filling with ultra-pure water. They went back on-line in 2006 as SuperKamiokande II.

hydrogen: D_2O. SNO is sensitive to all flavors of neutrinos, not just the type recorded by the Homestake and Kamiokande experiments, so was crucial in showing that neutrinos sometimes change from one flavor to another, important for such studies. SNO collected data through 2006, and is being updated as SNO+.

The neutrino problem

In all of these experiments the answer is the same: not enough neutrinos are detected. Typically, each experiment detects only about a fraction as many as should be seen. Originally, when the Homestake and Kamiokande results became known, it was thought that the theory could be adjusted, since those experiments are sensitive mainly to the much rarer solar reactions described in Appendix II (although some details of the comparison between Homestake and Kamiokande would still be very hard to explain). With the newer experiments, the situation deteriorated, since the shortage of low-energy neutrinos, those produced by the main fusion reactions and therefore tied directly to the total energy output, was even harder to explain.

This persistent shortage of solar neutrinos has come to be known as "the solar-neutrino problem." It is a "problem" only in the sense that there was until recently no consensus as to the cause. There are many possible explanations, ranging from the prosaic to the wildly exotic. For instance, models that tried to explain the low neutrino rate via something about the Sun generally seek to cool the interior, so that the balance of nuclear fusion processes is such that the same total energy is produced but the number of neutrinos is lower. Such solutions invariably run into the problem that the models of the solar interior are so good that most of these proposed answers can be ruled out.

One of the more exotic proposals, which has turned out to hold the answer, involves "neutrino oscillations," in which a quantum-mechanical effect (named MSW after the scientists who proposed it: Mikheev, Smirnov, and Wolfenstein) changes the neutrinos from the type that would be detected into another type, which would not have been detected with the existing experiments. There are three different types of neutrinos and the ones produced by the Sun transform into the other two types on their way to Earth. This surprising behavior was confirmed by the Sudbury Neutrino Observatory (SNO) in 2001. The result is a factor of 3 reduction in the number of neutrinos detected – exactly what the Homestake experiment showed.

Helioseismology

Any modern geology textbook will have a description of the Earth's interior, even though direct exploration of the layers beneath the Earth's surface has reached a depth of only a fraction of 1% of the Earth's diameter. How then is it possible for us to know the composition and structure of the interior?

One of the tools used in such studies is seismology. When a sudden large release of energy happens in a localized spot near the surface – such as an earthquake or an atomic bomb blast – the energy release causes a deformation of the Earth's crust that results in waves propagating outward, much like a hammer striking a gong. Some of these waves travel along the surface, while some move down into the interior, gradually spreading and weakening as they move away from the source.

Sensitive motion-detection apparatus located thousands of miles away can detect these disturbances, and can be used to determine the location and strength of the event. More important for our purposes, detailed analysis of the incoming waves at the detection site can provide information about the composition of the earth in between the source and the detector. Waves that started out moving downward into the earth are often deflected by boundaries or changes in composition, and may be bent back up to the surface. In this way it is possible through seismology to study the inner portions of the Earth that are not otherwise accessible. For instance, a certain type of wave called a "shear wave" cannot propagate through a liquid. So if a comparison of the waves reaching the detector via the Earth's surface with waves arriving after having travelled below the surface shows that shear waves are missing from the latter, then we know that the waves travelled through molten or liquid material on their way from the source to the detector.

In order to have wave motion two things are required: a disturbance and a restoring force. For instance, a block of gelatin sitting on your dessert plate has an equilibrium shape, more or less like a cube. If you poke it with your fork, you displace it away from its relaxed shape (this is the disturbance) while the attraction of the gelatin molecules to each other tends to restore the cube to its original shape. The result is that waves travel rapidly up and down the slab of gelatin, which appears to our slow eyes to be vibrating back and forth. A more traditional example is the plucked violin string. The string is stretched tightly between two attachment points, so that there is a tension force along the length of the string, which tends to keep it straight between the two ends. If the string is disturbed sideways (i.e., plucked) the tension will bring it back toward its preferred straight shape.

The alert reader might at this point ask why the gelatin or violin string goes through many oscillations, instead of merely returning directly to the undisturbed state. The answer is inertia. The gelatin or the string is not zero-mass, so that as it swings back toward the neutral condition, the speed of the moving matter pushes it right through the equilibrium position. The behavior is like a pendulum, which after having been displaced and released will swing through the vertical and up the other side, then back again. In principle, any of these oscillations would go on forever, unless the motions are eventually damped out. For instance, if the above experiments are carried out in a large vat of honey, it is clear that the gelatin or the violin string would quickly stop vibrating.

The surface of the Sun is seen to have oscillations, rather like the surface of a swimming pool. These oscillations turn out to be due to the combined effect of an enormous number of different waves running across and into the Sun, travelling in all directions and with a large range of sizes (or wavelengths); it is not just the surface, but the interior as well that is oscillating, as shown in Plate IXa. Most important for our purposes is that some of these waves penetrate deeply into the Sun and then return to the surface, as shown in Plate IXb. The wave reflects back from the surface because of the abrupt upper boundary at the solar surface, so they go down again, are bent up again, and so on. A wave may make many such bounces as it travels around the Sun.

What all of this means is that there are waves of various frequencies (and various wavelengths) that are reflecting up and down between the surface and a certain depth as they travel around inside the Sun. By very carefully measuring the properties of oscillations at the surface, which we can see, it therefore ought in principle to be possible to determine some properties of the solar interior, which we cannot see.

The figure shows that it is the longest wavelengths – those having the lowest frequency of oscillation – that penetrate most deeply inside the Sun. If we want to study the deep interior, we therefore need to have uninterrupted observations for a long period of time, in order to record enough full oscillations to get an accurate measurement. Unfortunately, we have day/night cycles at any given spot on the Earth, while measurements lasting several months or longer are needed. Several options are available. We can set up equipment at the South Pole during their mid-summer; this is difficult, expensive, and it also happens that the sky is often hazy down there. Another possibility is to set up a series of observing stations around the world, spaced so that continuous observing is possible. This is being done via the Global Oscillations Network Group (GONG). The other possibility is to put a telescope in space, at a location that does not have day/night cycles. This has been done on the Solar and Heliospheric Observatory, with its Solar Oscillations

Investigation experiment, and with the Solar Dynamics Observatory, with its Helioseismic and Magnetic Imager. The use of spacecraft is in many ways the best alternative; it is also the most expensive.

Now suppose that we want to determine the flow velocity of the sun at various depths. This information is actually crucially important for dynamo, or magnetic field generation, models of the solar cycle. Dynamo models are theories that explain the generation of magnetic fields inside the Sun by flow patterns of the hot interior gas, and progress in this field has been severely hindered by lack of knowledge of the subsurface flow patterns.

How do observations of wave frequencies at the surface tell us anything about flows inside the Sun? Imagine that there is a layer below the surface that is flowing from east to west at some speed. Now suppose that there is a wave of a certain frequency that travels down to that depth before being bent back up to the surface. If we are looking at a certain place on the surface, we will see that wave as it reemerges from below. Moreover, we will see that wave whether it originated east of our location or west of it. But if the wave originated west of our location, it had to flow against the subsurface stream to get to us; if it started east of us, then it flowed with the stream. This produces a difference in the frequency of the wave when it rises to the surface, depending on whether it was slowed down or speeded up by its subsurface travel. The result is a type of "frequency splitting" that provides a way of measuring the flow velocities below the surface, and measurements of the different types of waves will probe different depths.

A map of the solar interior flow speeds made by this technique is shown in Figure 4.4. This figure shows the rotation rate of the Sun at the surface (far right-hand part of the figure) down into the interior of the Sun (left-hand parts of the figure). This figure shows that the interior of the Sun rotates very differently

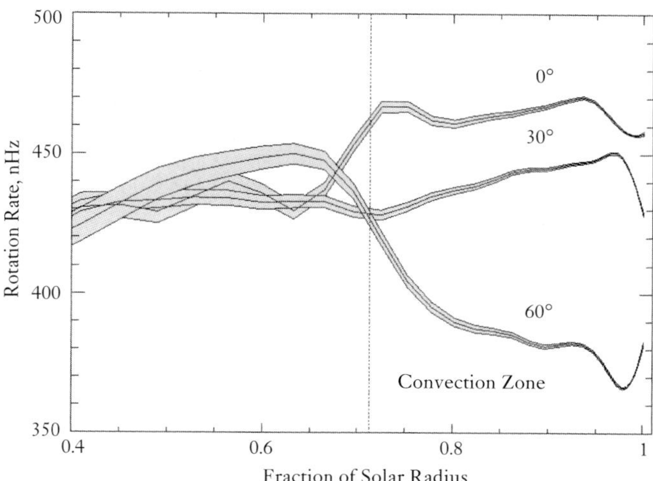

FIGURE 4.4. The pattern of flows inside the Sun, determined from helioseismology. The surface of the Sun has differential rotation, with high latitudes rotating more slowly than equatorial ones, whereas the core of the Sun rotates as a solid body.

than does the surface. The surface exhibits what is called "differential rotation." An ordinary globe rotates as a solid body: when you spin the globe, all of its parts take the same amount of time to complete a rotation. This is not the case for the Sun (nor for Jupiter, which also shows differential rotation). Parts of the solar surface near the equator rotate faster than parts near the pole. The right-hand portion of Figure 4.4 shows this effect (although the units need some explaining) with the Equator, at 0° rotating faster than the parts of the surface at 30° latitude, which in turn rotate faster than the parts at 60° latitude. The unit of measurement is given as "nHz," which is to say billionths of a cycle per second. Fewer cycles per second means slower rotation, and more means faster. The rotation rate is roughly 450 nHz, meaning that the Sun completes about 450 billionths of a cycle per second. This is another way of saying that the number of seconds it will take to complete one cycle is about 1 divided by 450 billionths, or a bit over 2 million seconds. This is roughly 26 days, which is indeed the length of time needed for the Sun's equator to complete a single rotation, as viewed from the Earth. But high latitudes are different, rotating at a rate of about 380 nHz, meaning that they require more than 30 days to complete a rotation.

Note, though, that the three curves come together and essentially coincide when we look at the interior of the Sun. When the different latitudes are all rotating at the same rate, this means that the Sun is rotating as a solid body, just as your globe does. So there is a change from differential rotation to solid-body rotation deep below the visible surface, at a distance of about 70% out from the center of the Sun (the change happens at roughly $r/R = 0.7$). This change, from solid to differential rotation was one of the predictions of the theory of the solar interior. Helioseismology therefore is able to confirm, in stunning fashion, the theoretical predictions, and also to provide details of the rotation patterns in the portion of the Sun that is believed to be responsible for producing the solar cycle. The method is sensitive enough to detect another component of the flow, a slow circulation at a speed of only meters per second from equator to poles and down again, that is crucial for resetting the field to begin a new cycle from the remnant fields of the old one. The method has even shown that there are at least two such cells of meridional circulation at different depths in the solar convection zone.

Once these interior flow patterns are known, the next step is to construct a dynamo model to generate the Sun's magnetic fields. Such a model would be a three-dimensional simulation which would, if all goes well, reproduce the main features of the solar cycle: the 11-year oscillation in sunspot number, the reversal of the magnetic field direction in alternate 11-year cycles, the migration of sunspots from high to low solar latitudes as the cycle progresses, the start of a new cycle as the old one is dying away, and so on. This work is just beginning, so it will be at least a few years before we know whether the mystery of the Sun's activity cycle has been solved.

A related use of helioseismology by the SOHO MDI and the SDO HMI teams takes advantage of the fact that strong magnetic fields at or below the surface lower the temperature of the Sun (this is why sunspots appear dark) and this in turn changes the sound speed. With good enough data, one can see magnetic flux concentrations below the surface, before they emerge (Plate X). With this technique it is also possible to determine the presence of large sunspots on the other side of the Sun, before they rotate around to become visible to Earth-based telescopes. Both of these methods offer the possibility of extending our predictive and early-warning capabilities for outbreaks of activity. We can also now see around the far side of the Sun with NASA's STEREO spacecraft (Chapter 6).

SUGGESTIONS FOR FURTHER READING

Chaplin, William J., *The Music of the Sun: The Story of Helioseismology* (Oneworld Publications, 2006).

Golub, Leon, and Pasachoff, Jay M., *The Solar Corona*, 2nd ed. (Cambridge University Press, 2010).

Kenneth J. H. Phillips, *Guide to the Sun* (Cambridge University Press, 1992).

Stix, Michael, *The Sun: An Introduction* (Springer, 2012).

Zirker, Jack B., *Sunquakes: Probing the Interior of the Sun* (The Johns Hopkins University Press, 2004).

5

Eclipses

The everyday sun dazzles the eye on a clear day. But about every year and a half, millions of people who are lucky enough to be in the right place on Earth see the brilliance of the Sun covered up. When even a single per cent is left visible, the sky remains blue and the event is not very spectacular. But when that last per cent disappears, the light level drops by an additional factor of 10,000, the sky turns black with pinkish color all around the horizon, the birds go home to roost and, as observers throughout the centuries have described, "day turns to night."

OBSERVING ECLIPSES

A total solar eclipse is an astounding sight, one that seems to awaken primal fears. Those who see one never forget it, and often come back to see more. The world of travel has advanced, and it is much easier now to fly off, as one of us did, to Siberia in 2008, to China in 2009, to Easter Island in 2010, to Australia in 2012, and to Gabon in 2013, than it was for scientists of 1936 to travel in a private railway car to Siberia, or even for Thomas Alva

Edison and other scientists in 1879 to travel from the eastern United States out to Iowa.

There were 66 total solar eclipses in the 20th century, so one every year-and-a-half is a solid average rate. It turns out there is about the same rate of annular eclipses, when a ring – or annulus – of everyday sunlight remains visible around the disk of the Moon. And there is an additional set of partial eclipses a year. Some years have only two solar eclipses of any type while other years boast of as many as five.

The cause of eclipses

Eclipses of the Sun occur when the Moon blocks our view of the Sun. Thus the Earth-Moon-Sun form a straight line. Such an alignment is a form of "syzygy," which is a wonderful word for an alignment of three celestial objects.

The Moon goes through its set of phases every $29\frac{1}{2}$ days, as it goes around the Earth while the Earth is going around the Sun. But the Moon's orbit around the Earth is tilted by about 5 degrees from the Earth's orbit around the Sun. So during most months, the Moon is either a little above or a little below the Sun when viewed from Earth at the phase we call "new moon." At that time, the far side of the Moon is fully lighted by the Sun, and we see only the near side, which is then the dark side.

But several times a year, the Moon is not far in its orbit from one of the "nodes," the points where the Moon's orbit around the Earth crosses the Earth's orbit around the Sun (Fig. 5.1). Then the Moon goes partially or fully in front of the Sun, and we have a solar eclipse.

The Moon's orbit around the Earth is elliptical. (The Earth's orbit around the Sun is elliptical too, but isn't as far out of round). So the Moon is sometimes closer to the Earth than at other times. At the times when it is closer, it appears slightly larger in angle on the sky. When this occurs, and the Moon

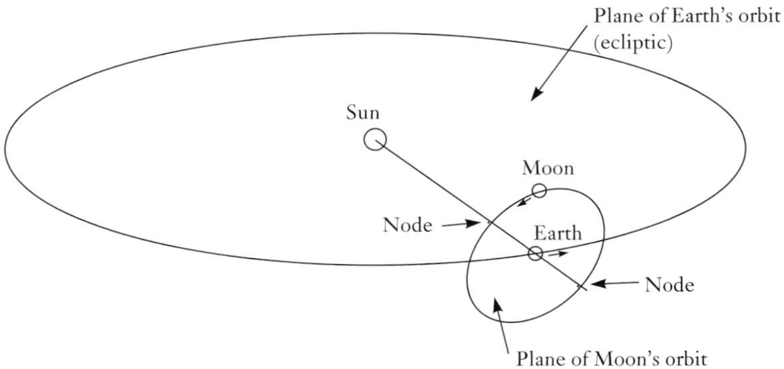

FIGURE 5.1. Relationship between Earth's orbit around the Sun and Moon's orbit around Earth, at a time when the nodes of intersection between these two planes give solar eclipses.

goes in front of the Sun so that their centers are close enough together, we have a total solar eclipse.

At this time, the shadow of the Moon falls upon the Earth. Since the Sun is larger than the Moon, this shadow tapers gradually (Fig. 5.2). As the Sun, Moon, and Earth move in their orbits, the Moon's shadow moves across the Earth. It can reach speeds of thousands of miles per hour. But the Earth rotates at the same time, and somebody on the Earth's surface keeps up with the Moon's shadow to some extent. The greatest speed of rotation on Earth is for an observer at the Earth's equator; after all, someone at one of the poles doesn't move at all as the Earth rotates. So the longest possible eclipse is when the Moon is closest to the Earth (making its shadow larger) and hits the Earth near its equator. Then a total solar eclipse can be as long as 7 minutes. The total solar eclipse of 1973 reached 7 minutes and four seconds in the Sahara Desert north of Timbuktu. Other eclipses are shorter, with eclipses of a minute or two or three being not uncommon. Still other eclipses may last only a second or so.

FIGURE 5.2. A total solar eclipse, drawn to the correct scale.

Thus a total solar eclipse occurs when the Moon is closest to the Earth. On the other hand, an annular solar eclipse occurs when the Moon is farthest, in its elliptical orbit, from the Earth and the point of the Moon's conical shadow doesn't quite reach the Earth's surface. Annular eclipses can last over 10 minutes. But perhaps the most exciting kind of annular eclipse lasts only a few seconds, as did one visible from Western Australia in 1999, or did the beginning of the annular-total eclipse of 2013 over the Atlantic Ocean. The mountains on the edge of the Moon may stick up enough so that an unbroken annulus is never visible, and the rapid variations in lighting can be quite dramatic.

The saros

The apparent sizes of the Sun and Moon are not the only happy coincidence that make eclipses so interesting. It also turns out that the circumstances of an eclipse – the precise geometric relationships of the Sun-Moon-Earth system – repeat 18 years $11\frac{1}{3}$ days later. Thus if you have a long eclipse near the equator, you will have another long eclipse near the equator 18 years $11\frac{1}{3}$ days later (or perhaps $10\frac{1}{3}$ or $12\frac{1}{3}$ days, depending on how the leap years fall). The long total solar eclipse of July 11, 1991, was followed by another long total solar eclipse on July 22, 2009, and we can confidently look forward to another wonderfully long eclipse, also with a peak duration over 6 minutes, on August 2, 2027.

This set of coincidences is known as the "saros," an interval discovered by the Babylonians thousands of years ago and named about three hundred years ago by Edmond Halley. The

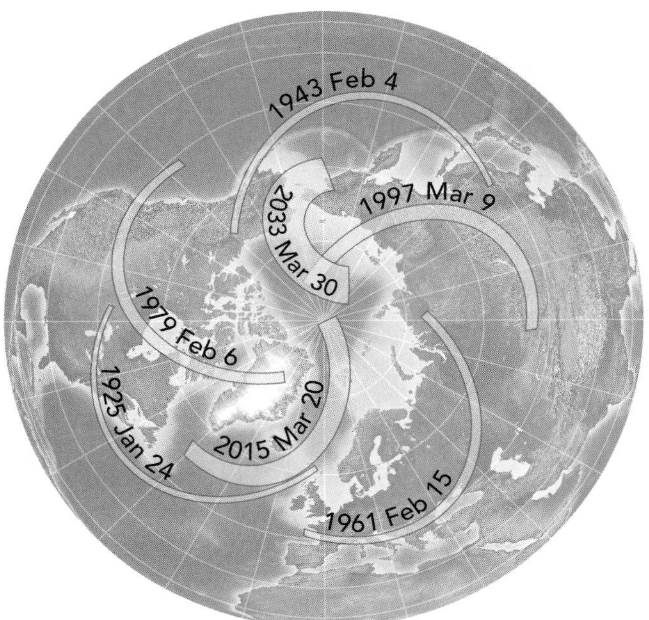

FIGURE 5.3. The final seven umbral solar eclipses of saros series 120, showing the characteristic saros pattern where each eclipse path is displaced 120 degrees west of its predecessor.

$\frac{1}{3}$ day in the saros has an interesting consequence: it allows the Earth to turn by $\frac{1}{3}$ of the way. Thus the eclipses in a given saros series have to be watched from different places on Earth (Fig. 5.3).

If you were out in space looking down at the Moon revolving around the Earth, you might think that it revolved in $27\frac{1}{3}$ days, which is how long it takes to be in the same position with respect to the stars. But the Earth's motion around the Sun means that the Moon has to rotate a bit further to again get in line between the two, which raises the interval to $29\frac{1}{2}$ days. This period is known as the "synodic period" of the Moon. One of these periods marks a "synodic month." After all, the word "month" came from the orbit of the Moon, and is a form of "moonth."

But we can measure the Moon's progress in its orbit in other ways as well. For example, we can measure how long it takes for the Moon to go from one node in its orbit to the next, where the nodes are the places where the Moon's inclined orbit crosses the Earth's orbit. This period is known as the "nodical month." Now, an eclipse occurs whenever the Moon is sufficiently close to its node, so these nodical months help define when eclipses occur. They are therefore also known as "draconic months," after the dragon who, in ancient Chinese myths, devoured the Sun to make an eclipse.

As it happens, 223 synodic months is almost exactly equal to 242 nodical months: 6585.32 days vs. 6585.36 days. Thus the Moon is coming back to its new moon phase after this period, and if it was near the node the last time it will be near the node this time too. Thus we have the saros of 6585.33 days, which works out to 18 years $11\frac{1}{3}$ days (plus leap-year effects). The Earth has gone around the Sun almost exactly 19 times with respect to the nodes in this interval, making 19 "eclipse years" also match the saros.

In yet a further coincidence, the period that it takes the Moon to go through its elliptical orbit around the Earth, varying in distance from the Earth, matches up as well. Some 239 of these "anomalistic months" equals 6585.54 days, nearly the saros defined by the synodic and nodical months. Thus 18 years $11\frac{1}{3}$ days later, you not only have another eclipse but you also have an eclipse with the Moon approximately the same angular size in the sky. This effect means that the duration of this successor eclipse is also similar. The 1973 eclipse peaked at 7 minutes 4 seconds; the 1991 eclipse peaked at 6 minutes 53 seconds; and the 2009 eclipse peaked at 6 minutes 39 seconds. Totality at the August 2, 2027, eclipse will last almost 6 minutes 22 seconds.

The saros is not perfect, and the Moon drifts slightly south-to-north or north-to-south over the years. A whole saros series lasts up to about 1500 years from the time when the Moon's shadow hits one of the Earth's poles to the time when the Moon's shadow drifts off the other pole.

Seeing an eclipse

Astronomers know the positions and motions of objects in the solar system so well that they can predict eclipses for thousand of years in the past and future. Over a period of a few years, they can predict the duration and timing of an eclipse for any place on Earth to an accuracy of about a second. The largest contribution to the inaccuracy comes from the shape of the Moon and the knowledge of which valleys or mountains will be on the Moon's edge as seen from Earth at the moments that totality begins or ends. (The detailed topographic mapping of the Moon with the Japanese Kaguya mission and the NASA Lunar Reconnaissance Orbiter now allow reconstruction of the details of the shape of the edge of the lunar silhouette, which leads to what, as we see below, are called Baily's beads.) The shadow of the Moon in space is a cone, and the Earth's curved surface intersects that cone in a complicated shape that is well approximated by an ellipse. Thus you can think of the Moon's shadow as an ellipse sweeping across the Earth at a speed of thousands of miles per hour. If you are within this cone, you will see a total eclipse. Since this darkest part of the cone is known as its "umbra" (from the Latin word for "shadow"), a total eclipse is an umbral eclipse.

If you are close enough yet off to the side of this elliptical shadow, either because it hasn't reached you yet or has passed you or because you are separated in a distance perpendicular to that of the path, you will see a partial eclipse: the Sun will appear only partially covered by the Moon. You are then seeing a partial eclipse (or the partial phases of a total or annular eclipse), which is a "penumbral eclipse," from the word for the outer part of a shadow.

The instant that the disk of the Moon first impinges on the disk of the Sun is known as "first contact." It is always a relief to detect first contact, since it indicates that you are in the right place and that you have not misread the predictions. The Moon then seems to gradually cover the Sun for an hour or two. For most of that time, the eclipse is not noticeable to those on the ground, since the remaining Sun is very bright, and our visual system is built to be relatively insensitive to variations in the brightness of the Sun, from clouds or haze. But as we approach "second contact," when the Moon first completely covers the Sun, the situation becomes more exciting. A few minutes before second contact, it grows noticeably darker; the wind may come up and the temperature will drop. The shadows look eerie and the quality of the light changes. Indeed, we are normally used to seeing our shadows look a bit fuzzy; though we don't think about it, our shadows have narrow penumbras around the darker, umbral portions. When only a sliver of sunlight remains, the shadows appear sharper than usual and, without necessarily realizing why, we notice that they look strange.

With a couple of minutes remaining, you may notice "shadow bands," subtle gradations of dark and light, sweep across the ground or across nearby walls. Some people even lay out a bed-sheet on the ground, so as better to see the shadow bands. These shadow bands undoubtedly are caused by moving regions of air in the Earth's upper atmosphere affecting the sunlight from a narrow sliver of Sun. With the Sun fully visible, the solar image is too broad to produce clear shadow bands. Atmospheric scientists may be interested in shadow bands, but professional astronomers are more excited about waiting for the next stage.

Seconds before the Moon entirely covers the Sun, the mountains on the leading edge of the Moon stick up through the thin sliver of sunlight remaining, breaking the sunlight into bits. These bits are known as Baily's beads, after the British 19th-century scientist Francis Baily, though they had been

FIGURE 5.4. The diamond ring effect and an adjacent additional Baily's bead, ending the 2012 total eclipse, photographed from a helicopter over Queensland, Australia.

seen previously by Halley and others. People around you are probably shouting for joy by the time Baily's beads are seen.

As the Moon continues to advance across the face of the Sun, the Baily's beads are extinguished. The last Baily's bead seems to glow so brightly that it looks like a diamond on a ring, and is known as the diamond-ring effect (Fig 5.4). This description was first made at the 1925 eclipse that crossed New York City, with crowds gathering on street corners to mark where in the grid of streets was the dividing line between total and partial eclipse. The diamond ring effect commonly lasts 5 seconds or so, but if the sky is especially clear and a deep valley is aligned at the edge of the Moon, the diamond ring effect can last for up to half a minute. If the sky is sufficiently clear, the circular band of the diamond ring may be formed by the corona.

A reddish edge to the Sun becomes visible. It got its name, "chromosphere," from its colorful appearance. It remains visible for only seconds, though its "flash spectrum," described in Chapter 3, reveals many of its secrets.

Then, the diamond ring is extinguished, marking "second contact." Totality begins. The crowd around you may cheer. Cameras will click away, even those of people who know better, in that they would have more fun in watching the eclipse in the sky instead of looking down at their cameras. Totality is the part of the eclipse most in use by professional astronomers, who want to study the corona that is then uniquely well visible from the ground. Though some aspects of the corona can be studied from certain high mountain tops with coronagraphs, the eclipse view is far superior. The eclipse studies from the ground are now increasingly linked with space observations of the corona, as we shall see later on.

Edmond Halley described the corona, at the 1715 eclipse he observed from England, as colored like a "pearl," and the description of the corona as pearly white remains common. The corona has a spiky appearance, with streamers coming out of the Sun's equator and thinner streamers known as polar plumes appearing nearer the poles. These beautiful structures are made of hot coronal gas held in place by the Sun's magnetic field. The shape of the corona varies over the solar-activity cycle, ranging from relatively round at solar maximum, when streamers appear at all latitudes on the Sun, to relatively oblong at solar minimum, when streamers appear only near the Sun's equator.

During totality, the sky is dark in the direction of the Sun and Moon, since you are then in the Moon's shadow. The planets Venus and Mercury become visible, if they are above the horizon, and other planets and stars may be visible as well. Far from the eclipse in the sky, light is scattered into your view from regions outside the eclipse. So how dark the sky actually becomes depends on how wide the eclipse path is and on how clear the sky is. Usually, the horizon looks reddish all around, like a 360 degree sunset. As Halley wrote, "From this time the Eclipse advanced ... when the Face and Colour of the Sky began to change from perfect serene azure blew, to a more dusky livid

Colour having an eye of Purple intermixt, and grew darker and darker till the total Immersion of Sun...."

After all too short a time, seconds or minutes, the chromosphere and a second diamond ring appear, marking "third contact." Baily's beads brighten the sky within seconds, and the exciting part of the eclipse is over. Hardly anybody bothers to watch the final partial phases, which end an hour or two later with "fourth contact."

Imaging solar eclipses

Though the most exciting thing may well be to watch a total eclipse directly with your eyes, most of us succumb to the temptation to photograph it. All too many people travel to the eclipse zone and then never actually see the eclipse because they are fiddling with equipment. Still, it is fun to have your own photos.

Most cameras have lenses too short to make good closeup pictures of the corona. The size of the sun and moon on your film depend on the focal length of your lens. As a rough rule, a 500-mm lens gives an image about 5 mm across for the Moon or for the solar photosphere on 35-mm film. So if you add an equal amount of corona on either side, such a lens fills most of the frame with the eclipse for a "full-frame" digital camera that matches the sensor size of 35-mm film. Most digital cameras' sensors are only about $\frac{2}{3}$ that size, so a 400-mm lens would be appropriate for the corona during eclipse. You must have your camera on a sturdy tripod to keep it from shaking, and using a remote shutter release so that you are not physically touching the camera is also helpful (Fig. 5.5).

If you are using more conventional lenses, like 50-mm standard lenses on a 35-mm camera, then the image of the eclipse itself will be very small on the film. You may find it best to use the eclipse as a design element in the photo rather than as the central element. Put some people or objects in the foreground.

FIGURE 5.5. Photo of the solar-maximum corona at the 2013 African total eclipse, taken using a 400-mm-focal-length telephoto lens on a full-frame camera from Gabon.

Just how wide a field of view you need depends on how high the eclipse is in the sky, or on how low you can crouch (Fig. 5.6).

The corona is brightest near the solar limb and falls off rapidly with distance outside the limb. In fact, it is over 100 times brighter near the limb than it is only one solar radius out. (Solar astronomers often talk in solar radii, merely meaning half the angular diameter of the Sun in the sky.) We would like to record the corona out to at least several radii and film doesn't easily cover that much range in brightness, so photos will capture only a limited part of the corona at the proper exposure and will be overexposed or underexposed for other parts.

The American astronomer Gordon Newkirk developed a method in the 1960s for capturing more of the brightness range on a single piece of film. He and his colleagues, mostly from the High Altitude Observatory in Boulder, Colorado, made a filter that had absorbing and reflecting material on it, deposited in a shape that matched the overall average brightness of the

FIGURE 5.6. Photo of the corona at the 2012 total solar eclipse, taken using a wide-angle 10.5-mm-focal-length lens from a helicopter over Queensland, Australia. The umbral shadow of the Moon shows clearly.

corona. Thus his filter was densest in a ring just the size of the Moon, and became less and less dense outward in the radial direction. It was thus called a "radially graded filter," or, for short, a "radial filter." This filter was used in a device, commonly called a Newkirk camera, to make images of the corona at a series of eclipses (Fig. 5.7).

The standard Newkirk camera did more than merely take these pretty pictures; it also made observations of the polarization of the corona, and the observations were carefully calibrated so that scientists could study the density of electrons in the corona. By the 1998 eclipse, the original Newkirk camera was retired, and a camera using an electronic detector was used instead.

As techniques were developed for scanning film, it became possible to make composite photographs out of several individual ones, each exposed to show a different brightness region of the corona. The longer the exposure, the farther out in the

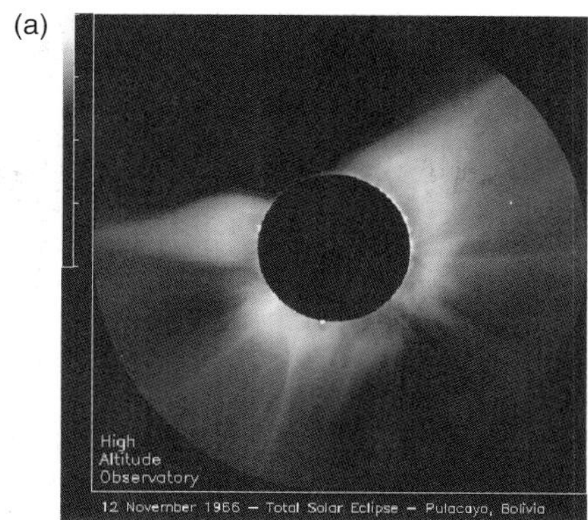

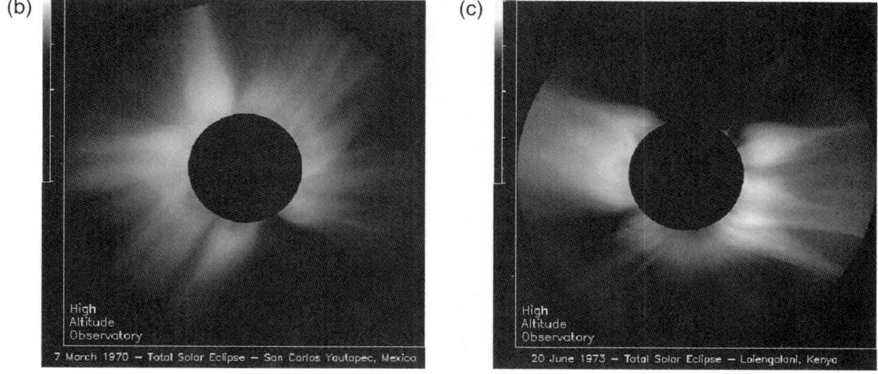

FIGURE 5.7. Solar eclipse images made with the Newkirk camera by the staff of the High Altitude Observatory of the U.S. National Center for Atmospheric Research, which is operated by the University Corporation for Atmospheric Research and sponsored by the National Science Foundation; the 1980 and 1991 expeditions were conducted jointly with Rhodes College. (a) Nov. 12, 1966; (b) March 7, 1970; (c) June 30, 1973; (d) Feb. 16, 1980; (e) July 31, 1981; (f) June 11, 1983; (g) March 18, 1988; (h) July 11, 1991; (i) Nov. 3, 1994; (j) Feb. 26, 1998. The last image uses a numerical processing method to replace the radially graded optical filter.

corona you see. In the 1990s, these techniques were perfected so much that composite images showed well the coronal structure of a wide range of solar radii outside the limb (Fig. 5.8).

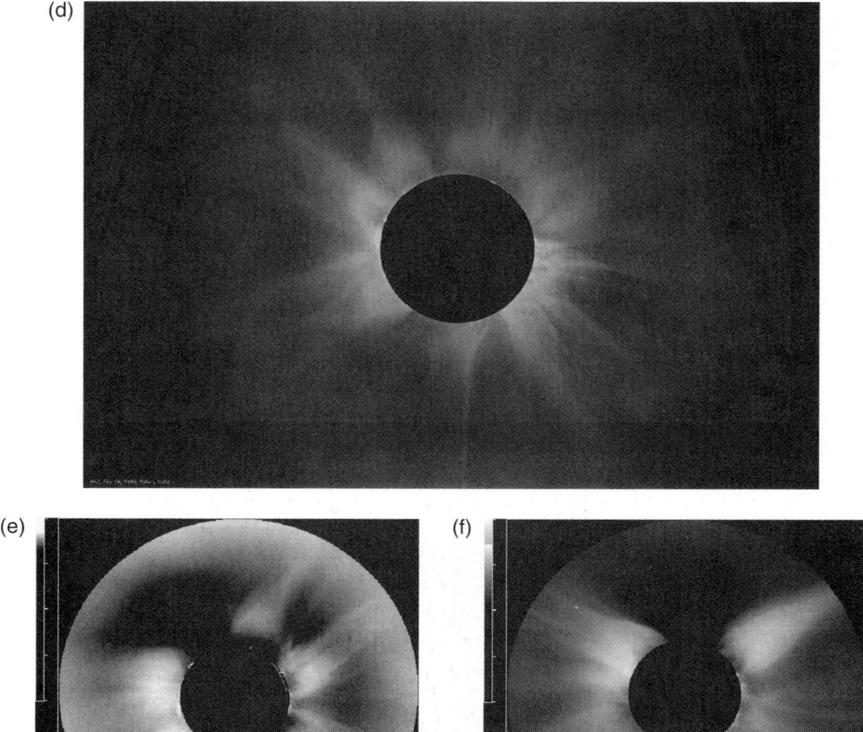

FIGURE 5.7. *(cont.)*

Through that decade, film became replaced by electronic detectors in much of astronomy, and eclipse photography was no exception. Most of the detectors are of a type called charge-coupled devices, or CCDs. Even the camcorder you may see advertised in your newspaper or online has a CCD in it, though the ones professional astronomers use are more carefully made as well as being more precisely calibrated and being more controllable in exposure time and other functions.

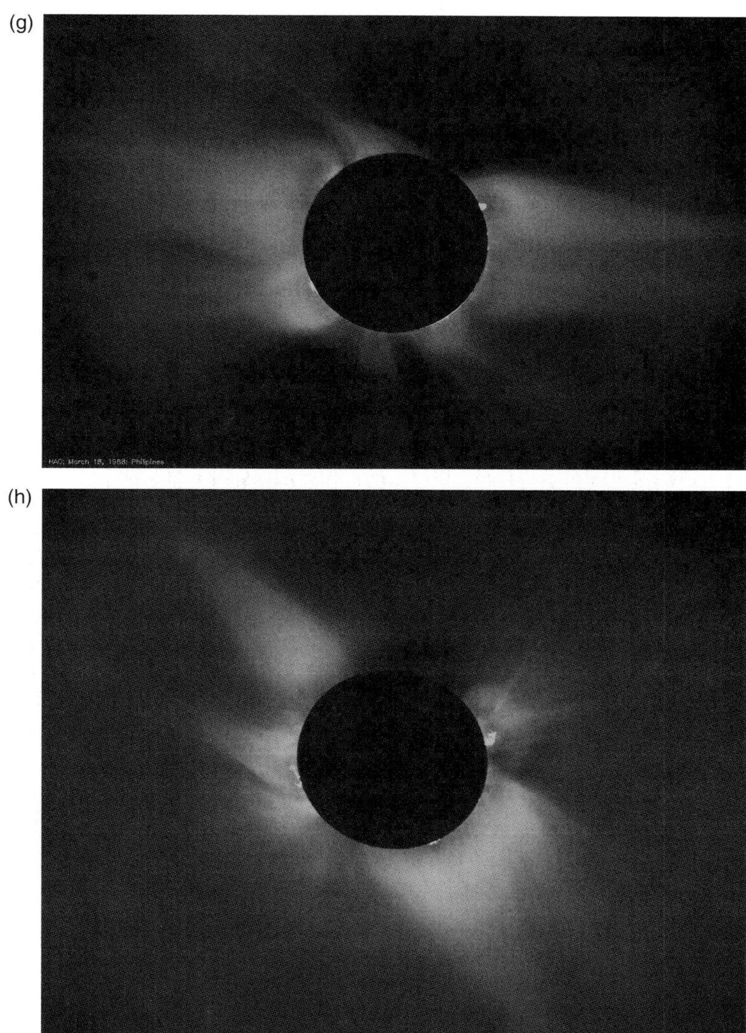

FIGURE 5.7. (*cont.*)

CMOS chips (Complementary Metal Oxide Semiconductor chips) are now also in use for cameras. Many people now take eclipse photos using merely their iPhones, though those are wide-angle views. (Some of us are organizing a “Megamovie”

(i)

(j)

FIGURE 5.7. *(cont.)*

(a)

FIGURE 5.8. Solar eclipse composite images made from dozens of original photographs combined to bring out the full range of bright and faint features in the corona. The computer work to composite many individual frames to make each of these composites was carried out by the New York amateur astronomer Wendy Carlos from individual frames taken at Williams College expeditions. (a) Aug. 11, 1999; (b) June 21, 2001; (c) Dec. 4, 2002; (d) March 29, 2006; (e) Aug. 1, 2008; (f) July 22, 2009; (g) July 1, 2010; (h) November 14, 2012.

project, http://www.eclipsemegamovie.org, with iPhones and other cameras for the U.S. eclipse of 2017.)

Many people now make videos of eclipses, using their CCD camcorders or the video function on their cameras or phones. Standard, consumer cameras usually have zoom lenses so powerful that the corona can be made to sufficiently fill the frame. A hint is to focus the camera manually in advance, since otherwise the automatic focus feature may "hunt" back and forth, losing valuable time in the all-too-brief total phase of the eclipse.

(b)

(c)

FIGURE 5.8. *(cont.)*

(d)

(e)

FIGURE 5.8. *(cont.)*

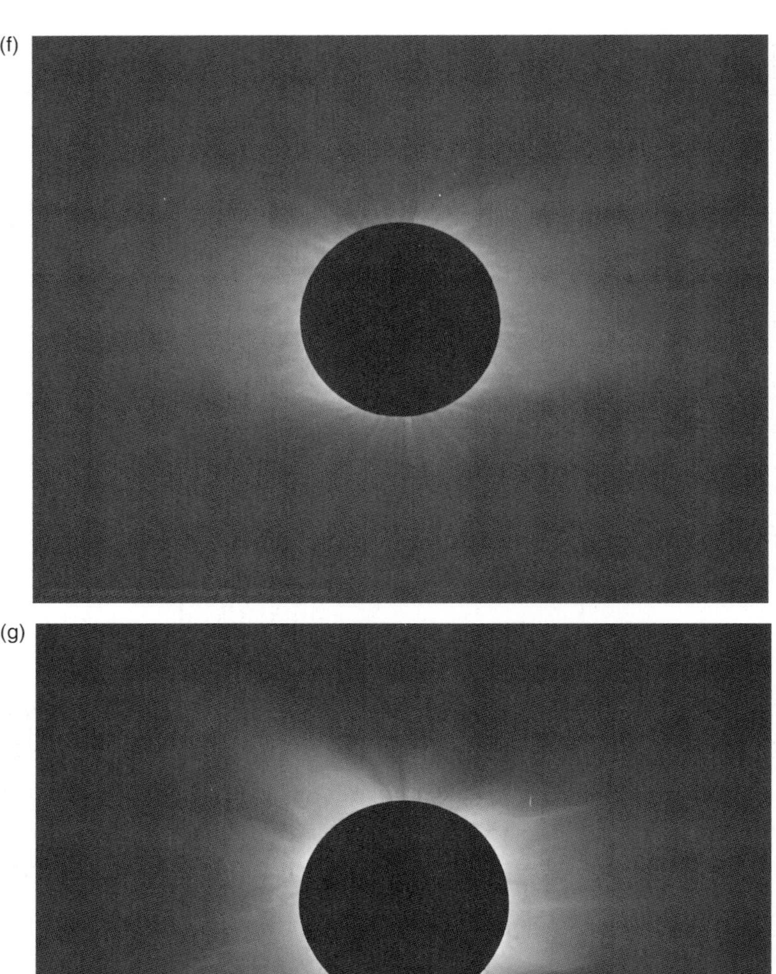

FIGURE 5.8. *(cont.)*

(h)

FIGURE 5.8. *(cont.)*

Note that only during totality, from diamond ring to diamond ring, can you look at the Sun directly or can the camera image the Sun directly. Before second contact or after third contact – that is, during the partial phases – you need to use special solar filters to reduce the intensity of the solar photosphere to safe levels for your eye or for your camera.

So the routine for taking photographs of an eclipse is to have the filters in front of your camera lens until the diamond ring appears, and then to take it off for the interval of totality. It must then go on when the second diamond ring appears. Just how well you handle that timing affects how good your eclipse video will be.

Eye safety at eclipses

All too many people think that eclipses are hazardous in some special way, perhaps because some special dangerous rays come out of the corona. That is not the case. The eclipse is a time when there is less light coming out of the sun, rather than more of anything. The corona is always there, but you can't normally see it because the photosphere is too bright.

The everyday solar surface – the photosphere – is too bright to look at safely, and as long as any of the photosphere remains visible, its surface brightness can cause damage to the eyes of people who stare at it or who look at it directly through an optical device like an unsatisfactorily filtered telescope or binocular. The eye alerts you to excessive total brightness, but it is fooled at an eclipse because only a tiny piece of very bright photosphere remains, so your eye does not sense the danger.

You can buy a special solar filter that reflects or absorbs all but one part in 100,000 of the incoming sunlight. The ones made out of polyester with a coating of aluminum or of a chrome/nickel alloy, or out of absorbing black polymer, can be found for about a dollar. Still, we use such filters only for a glance of a second or two every few minutes during the partial phases. There is no point on staring for longer at a partially eclipsed sun: it isn't fundamentally interesting and it is best to wait for totality, when you put your filters down and glory in the direct view of the solar corona.

The solar corona is about as bright as the full moon, and equally safe to look at. So for the period of totality, and only for that period, can you and should you look at it directly, without filters.

During the partial phases, you don't even have to use a solar filter to keep track of how the eclipse is going. You can make a simple pinhole camera instead. To do so, merely punch a hole a few millimeters across in a piece of paper or cardboard (or, to be a little better, in a piece of aluminum foil). Then hold that piece with the hole in it up in the air, and allow it to project an image of the Sun onto a piece of paper or cardboard held about two feet down toward the ground. You will see a crescent sun on that piece of paper, and since you are looking down at it, with the Sun at your back, it is completely safe to look at. Of course, you don't look up at the Sun through the hole.

Historical eclipses and the rotation of the Earth

Although astronomers can predict the positions of the Sun, Earth, and Moon thousands of years into the past or future, one confounding factor in predicting eclipses is that the rotation of the Earth is not regular enough to allow the exact location of the Moon's shadow on the Earth to be accurately known over periods of many millennia.

F. R. Stephenson of the University of Durham in England, and his colleagues, have surveyed historical reports of eclipses in various parts of the globe for thousands of years. From their work, they have found how the Earth's rotation has varied over time, gradually slowing down. Their results have importance for geology, since they reveal processes under the Earth's surface that affect the slowdown. For example, after the era of glaciers, uplift of the ground may have changed the Earth enough to affect its rotation in this way. Using observations over the past 2500 years it has been found that the dominant effect is a slowing of the rotation due to tidal friction, increasing the length of the day at an average rate of about 2.3 milliseconds per century. There is also a variation, perhaps due to post-glacial uplift, toward a shorter length of day by 0.6 milliseconds per century, giving a net effect of about +1.7 milliseconds per century. Thus over the last 2500 years, the day has increased in length by about 42 milliseconds, or about $\frac{1}{25}$ second.

Accurate prediction of both time and location of solar eclipses is a relatively recent phenomenon. The story is often related in books on eclipses how a total solar eclipse on May 28, 585 BC, predicted by the Greek astronomer Thales of Miletus, stopped a war between the Medes and the Lydians. The story was related by the historian Herodotus (c. 484-425 BC), who wrote about 150 years after the event. Herodotus wrote, "When the Lydians and the Medes saw the day turned to night, they ceased from fighting, and both were the more zealous to make peace." But Stephenson's evaluation is that the account is of "doubtful reliability," and that the methods of predicting eclipses were not far enough advanced in 585 BC for the story to be true. Stephenson concludes that "the assertions by Herodotus and Pliny probably owe their origin to the various legends which accumulated around the personality of Thales."

Future solar eclipses

The distribution of total eclipses across the Earth's surface is sporadic (Fig. 5.9). If you stand in one place, it may take centuries for a total eclipse to be visible overhead. Fortunately, it is relatively easy to travel these days, to take advantage of eclipses much more often than that.

On November 14, 2012, an eclipse occurred over Australia and then continued over the Pacific Ocean. The November 3, 2013, eclipse's totality (the eclipse's beginning was annular, though only over the Atlantic Ocean) crossed Africa from Gabon to Uganda, Kenya, Ethiopia, and Somalia in a narrow band.

On March 20, 2015, totality will be visible only in the Arctic, including Svalbard and the Faroe Islands. Partial phases will be visible throughout Russia west of Novosibirsk and the middle of Mongolia, and throughout Europe, northern Africa, and the northern Middle East. 65% of the Sun's diameter will be covered at Moscow, 78% covered at St. Petersburg, and 39% covered at Novosibirsk.

On March 9, 2016, totality will cross several Indonesian islands, from Sumatra to the northeast, including Kalimantan/Borneo and Celebes (but not Java).

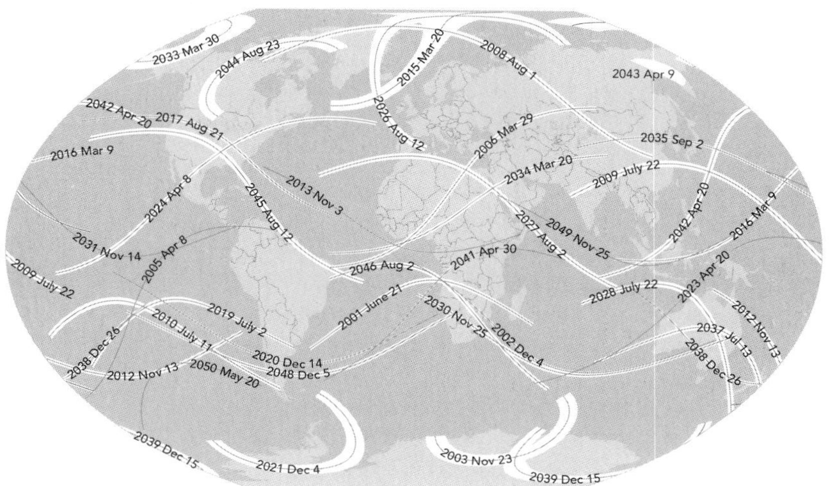

FIGURE 5.9. Total eclipse paths traced on the Earth's surface, for the years 2010–2060.

Many people are awaiting the total solar eclipse of August 21, 2017, at which totality will cross the continental United States, from northwest to southeast, with the whole Continental, Canada, Central America, and northern South America having partial phases.

On July 3, 2019, totality, after crossing the Pacific, will reach Chile, with some of the world's major nighttime telescopes as well as the ALMA millimeter-submillimeter array in the path, and with partial phases visible throughout all but northernmost South America.

The total solar eclipse of December 14, 2020, will cross Chile and Argentina, with partial phases visible throughout southern South America.

The total solar eclipse of December 4, 2021, will be visible only from Antarctica.

Totality at an annular-total ("hybrid") eclipse of April 20, 2023, will cross Australia and Indonesia.

Table 5.1 Total solar eclipses, 2013–2027

Date	Maximum totality (min:sec)	Locations
November 3, 2013	1:40	Atlantic, Africa: Gabon to Kenya, Ethiopia, Somalia
March 20, 2015	2:47	Svalbard, Faroe Islands
March 9, 2016	1:57	Indonesia, Pacific Ocean
August 21, 2017	2:40	Continental United States
July 3, 2019	4:33	Pacific, Chile, Argentina
December 14, 2020	2:10	Pacific, Chile, Argentina, Atlantic
December 4, 2021	1:54	Antarctica
April 20, 2023	1:16	Exmouth, Australia; East Timor and Irian Jaya, Indonesia
April 8, 2024	4:28	Mexico, United States
August 12, 2026	2:18	Spain, Iceland, Greenland, Russia west of Laptev Sea
August 2, 2027	6:22	Southermost Spain, Gibraltar, North Africa, Saudi Arabia, Yemen, Somalia

Totality on April 8, 2024, will cross Mexico and the United States.

Totality on August 12, 2026, will cross Spain, the north Atlantic Ocean, Iceland, and Greenland, pass near the north pole, and end in northernmost Russia near the Laptev Sea. An 84% partial eclipse will be visible at Moscow, with 83% at St. Petersburg.

An annular eclipse of June 1, 2030, will cross from north Africa through Greece and Turkey and then across southern Russia. 79% of the Sun's diameter will be covered at Moscow and 93% covered at Novosibirsk.

Annular eclipses aren't as spectacular as total eclipses, but can be fun to watch. Filters must stay on for the whole eclipse, including annularity.

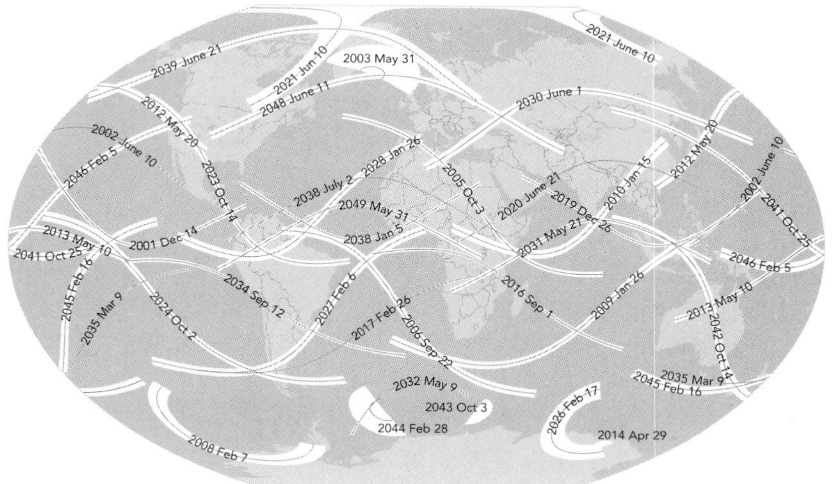

FIGURE 5.10. Annular eclipse paths traced on the Earth's surface, for the years 2010–2060.

The scientific value of eclipses

People often ask whether, in this age of space observations, there is still scientific value to observing eclipses. The answer is that surely yes, there are still many things to be learned from solar eclipse expeditions on Earth that cannot yet be found from space observations. And there are uses for making coordinated observations from the ground and from space.

Similarly, for non-solar astronomy, the question is often asked why ground-based telescopes are necessary now that the Hubble Space Telescope is in orbit. The answer for that question is also that Hubble is only one small telescope, and that there is lots to do separately from Hubble and also lots to do in following up problems suggested by Hubble observations.

The eclipse of 2010, for example, was the first eclipse when the x-ray telescope XRT on the Japanese Hinode spacecraft, the ultraviolet and visible-light telescope on the Solar and Heliospheric Observatory (SOHO), and the ultraviolet and visible-light telescopes on the Solar Dynamics Observatory (SDO) were all aloft, while imaging the solar corona from ground-based eclipse sites fills in some gaps in the coverage from space. For example, the coronagraphs on SOHO – and most other coronagraphs as well – have to occult (hide) not only the photosphere but also a region around the photosphere about half a solar radius in extent, in order to limit the amount of scattered photospheric light that contaminates the coronal image. It is exactly this inner corona region – missing from the space-based data – that is well imaged by ground-based eclipse observations, because the Moon provides an ideally-placed occulting disk having only minimal scattering and diffraction problems.

Combining ground-based observations at eclipses with the space-based coronagraphs also has the potential of improving knowledge of the characteristics of the light scattered around in the space telescope, improving the interpretation of data taken with it. Figure 5.11 shows such a combination, allowing scientists to follow the magnetic structure of the coronal regions all the way from the solar surface out into the open regions where radially oriented streamers dominate.

The Atmospheric Imaging Assembly on NASA's Solar Dynamics Observatory spacecraft makes exquisitely detailed observations of loops of hot coronal gas held in place by the Suns magnetic field. But the telescope observes at a cadence of about a second between exposures. So an experiment that images the Sun over and over again at 10 times per second, such as one of the eclipse experiments run by one of us, fills a part of observational space that is not otherwise covered.

The questions to study at a ground-based eclipse expedition must be carefully chosen to complement or to be separate from observations made from telescopes in space. But the ability to use new equipment, perhaps to study newly developed ideas, and to do so at a relatively low cost compared with space

FIGURE 5.11. The solar corona is so large in angular extent that we typically must use several different ways of recording it in order to show it from the photosphere out to the region where it consists mostly of radial streamers. This figure shows combined images obtained at the time of the November, 2012, total eclipse using an EUV imager (the SunWatcher, SWAP, 80% sized), on disk, a white-light eclipse image from Gabon for the inner corona, and the SOHO LASCO C2 coronagraph for the outer corona.

observations, leaves ground-based eclipse studies a role to play for the forseeable future.

Eclipse tourism

Of course, most people seeing solar eclipses are not professional astronomers. It is unusual to have a total eclipse pass near you, but most of the people seeing any individual eclipse are just those people who are near home. The 1999 eclipse was notable in that it crossed Europe in a path that was close to the homes of tens of millions of people. The 2017 eclipse path of totality will be within driving range for hundreds of millions of Americans.

Increasing numbers of tourists, many with no special interest in or background in astronomy, are travelling in groups to see eclipses. Eclipse cruises have become big business, with many ships carrying tourists to see the totalities, such as those in the Pacific east of Australia for the total eclipse of 2012 or in the Atlantic west of Africa for the total eclipse of 2013. Many land tours take people for two weeks or so to a variety of sights, making sure only to be in totality at the proper time on eclipse day.

THE CORONA

With the advent of photography in the 19th century, the corona became a familiar sight, especially when imaging techniques improved, as described above. But without photography, or some other means of recording the image, very few people would ever see a total solar eclipse, because the ribbon of area in full umbral shadow is so narrow. There is ample historical evidence that total eclipses of the Sun were known thousands of years ago, but it is unclear whether there is any historical evidence prior to the 17th century that anyone knew that the Sun has a corona. An example of how ambiguous the historical record can be is an alleged report of a Babylonian eclipse in 1063 BC:

> "On the twenty-sixth day of the month of Sivan in the seventh year the day was turned to night, and fire in the midst of the heaven."

This is the sort of record that would, a few years ago, have been taken to indicate not only a total eclipse (day turned to night) but a sighting of the corona as well (fire in the heaven). But today historians are more cautious. In analyzing this and other reports, the historian Roger Newton points out

> "If the passage refers to the eclipse of –1062, both the month and the day of the month are given wrong. The passage does not mention the Sun directly and it does not use the standard Babylonian term for an eclipse of the Sun. There are plausible explanations for all these discrepancies, but the combination of so many of them in one record does not inspire confidence."

One of the first clear mentions of something that sounds very much like the corona was made by Kepler, who specifically asked: "why does the Moon remain partly visible when the Sun is behind it during total eclipse?" Kepler, however, seemed to think that the corona is caused by scattering of sunlight in a lunar atmosphere: "... the whole Moon is surrounded by some sort of aery essence, which reflects rays from all parts."

This question of whether the glow around the Moon does or does not belong to the Sun was not settled until more than two centuries later. Since it had already become clear by then that Moon does not have an atmosphere, the lunar explanation had been ruled out. Could the glow around the Moon then be caused by the Earth's atmosphere? Photographs of the corona were taken during the eclipse of 1860 from widely separated sites. These showed that the features seen at both locations were the same, providing good evidence that the phenomenon was connected to the Sun. Then in 1869 a strong emission line coming from the corona was found in the green part of the spectrum, which was taken as conclusive evidence that the corona is solar – such an emission line could only come from a hot gas. But this discovery, in turn, raised a problem that took fifty years to resolve: why is the corona emitting this line?

The great coronium puzzle

Observations at the 1868 eclipse led to the discovery of a bright emission line in the spectrum of the chromosphere, which is normally not observable except for a few seconds just prior to and just following totality. What happened next is nicely described by C. A. Young in the 1895 edition of his book, *The Sun*:

> The famous D_3 line was first seen in 1868, when the spectroscope was for the first time directed upon a solar eclipse. Most observers supposed it to be the D line of sodium, but Janssen noted its non-coincidence; and very soon, when Lockyer and Frankland took up the study of the chromosphere spectrum, they found that the line could not be ascribed to hydrogen or to any then known terrestrial element. As a matter of convenient reference Frankland proposed for the unknown substance the provisional name of "helium" ...
>
> Naturally there has been much earnest searching after the hypothetical element, but until very recently wholly without success ... The matter remained a mystery until April, 1895, when Dr. Ramsey, who was Lord Rayleigh's chemical collaborator in the discovery of argon, in examining the gas liberated by heating a specimen of Norwegian cleveite, found in its spectrum the D_3 line, conspicuous and indubitable ... Cleveite is a species of uraninite or pitch blende, and it soon appeared that helium could be obtained from nearly all the uranium minerals ...

As we now know, the connection between uranium and helium is that radioactive decay of uranium involves what were at that time called "alpha" particles, which are helium nucleii. These nucleii pick up electrons to become atoms of helium, which can become trapped in uranium-rich rocks, to be released when the rocks are heated.

Meanwhile, in 1869, a year after the discovery of the helium line on the Sun, Young and W. Harkness, working independently, found a strong green coronal emission line that could not be identified (Fig. 5.12). As had been done the year before a new element, named "coronium," was proposed. To quote Young again, writing in 1895: "the name 'coronium' has been provisionally assigned to it, and the recent probable identification of

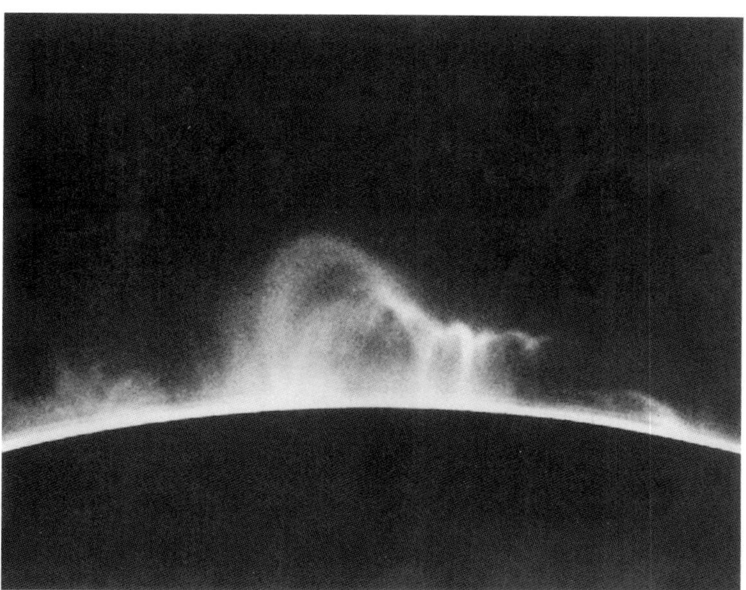

FIGURE 5.12. Images of coronal loops taken in the "coronium" green line.

'helium' in terrestrial minerals gives strong reason to hope that before very long we may find coronium also."

This new element had a most unusual property: it had to be much lighter than hydrogen, the lightest known element. The reason for this is that the corona – the Sun's atmosphere – is seen to extend very far out from the solar surface. The height of an atmosphere is determined by the balance between the downward pull of gravity and the upward mobility of the atmosphere's gas particles from their temperature. For a given temperature, lighter particles move more quickly than heavy particles. In essence, temperature is a certain amount of energy of motion per particle, and light ones need to move faster than heavy ones to have the same amount of energy. Like a ball tossed in the air, the particles move upward to some height – determined by their initial upward velocity and by the downward pull of gravity – then they fall back down. Averaging over all of the particles in the atmosphere, we find that the density of the gas falls off as we move upward, gradually becoming less and less.

So there is a quantity called the "scale height," which contains the ratio between the temperature of the gas and the mass of the particles. This ratio tells us how far upward the atmosphere will extend. For the corona, the scale height for any ordinary element, even hydrogen, at a photospheric temperature is so small that only a thin skin sitting just above the solar surface would be visible. Since the corona in reality extends upward to great distances, we are forced to conclude that coronium must be an extremely light element, much lighter than hydrogen.

This conclusion posed a tremendous problem, especially after Mendeleev's work on the periodic table of the elements. In 1871, Mendeleev showed how the elements form a systematic arrangement and it soon became clear that there was no room for coronium in the pattern. The choice was then to either question the periodic table, or to get rid of coronium. But then what was causing the extended coronal emission?

The fortunes of coronium were further damaged by Rutherford's work in 1911, showing that atoms consist of a small, dense nucleus surrounded by lightweight electrons. Simple counting implies that hydrogen, with only a single nucleon, is as light as an element can be. So now there was no room, even in principle, for coronium.

Many explanations for the mysterious coronal emission lines (several were known by then) were proposed in the early decades of the 20th century, but it was not until 1939 that the answer was finally found. It came through the convergence of many separate avenues of research, ranging from laboratory experiments on the spectra of spark discharges, to astronomical studies of nebulae (for which the element "nebulium" had been hypothesized). As often happens, once the answer is known, one feels that it should have been obvious all along: the corona is extremely hot.

The fall of coronium

The reason that the coronal emission lines were so hard to identify is that they arise from gas in a state that is not normally encountered, and therefore one that investigators had not experienced. As a gas is heated, the violent collisions between atoms in the gas become strong enough to knock electrons up into higher energy levels of the atom (a process called "excitation"), and sometimes to knock them out of the atom completely ("ionization"). These processes had been well studied throughout the 19th century, but for relatively low energy versions and mainly for ionization of the neutral atom, that is, the atom supplied initially with all of its electrons.

Studies were gradually progressing toward more exotic conditions, in which more than one electron is removed from an atom, and the prevailing conditions (such as high temperature and low density) are such that the atom remains in a multiply ionized state for a long time. This turned out to be the path toward explaining the coronal spectrum. The corona contains emission lines from such unfamiliar states as, for example, nine-times ionized iron, and thirteen-times ionized calcium. These are not at all obvious choices for explaining the observed spectral lines, and even once they were suggested, the calculations needed to predict the wavelength of, say, the coronal green line, are difficult.

Based on a long series of laboratory experiments by the Swedish spectroscopist Bengt Edlén and by Ira Bowen at Caltech, the "nebulium" emission lines were identified as coming from highly ionized states, such as five-times ionized iron. This work gave Walter Grotrian, in Potsdam, the clue to the coronium lines. He suggested that nine and ten times ionized iron would produce spectra that included rare atomic transitions at visible wavelengths. When Grotrian's brief note on the subject was published in 1939, Edlén looked over his laboratory work and noticed that his spark discharge spectra supported this view. Very quickly he determined that the famous green line could be explained by one of these rare optical transitions in thirteen times ionized iron. The rest, as they say, is history.

The hot corona

Once the breakthrough was made, the corona was once again a physically possible object, and the problem became one of explaining not what it is, but how it gets that way: why is the corona so hot? This is still one of the great outstanding puzzles in solar research today.

Early investigators did not immediately realize how hot the corona is because there are at least four different types of light coming to us, and only one of them is actual emission from the hot coronal gas (Figure 5.13). The others are due to scattering or reflection of the extremely bright photospheric light, as listed below:

K-corona From the German word "Kontinuierlich," or continuous, because it emits light over a broad spectrum, this is the brightest component of the corona in the visible. It arises out of photospheric light that is scattered by the electrons in the corona. The spectrum appears continuous, even though the photosphere has the strong absorption lines found by Fraunhofer, because the electrons are moving so quickly in random directions in the corona that the lines are smeared out, becoming so broad that they are no longer detectable as lines.

F-corona So-named because it shows the dark absorption lines found by Fraunhofer, this component is due to scattering of the photospheric light by grains of dust in the ecliptic plane. The solar system is filled with such dust, which also causes the faint glow at night known as the zodiacal light. Because

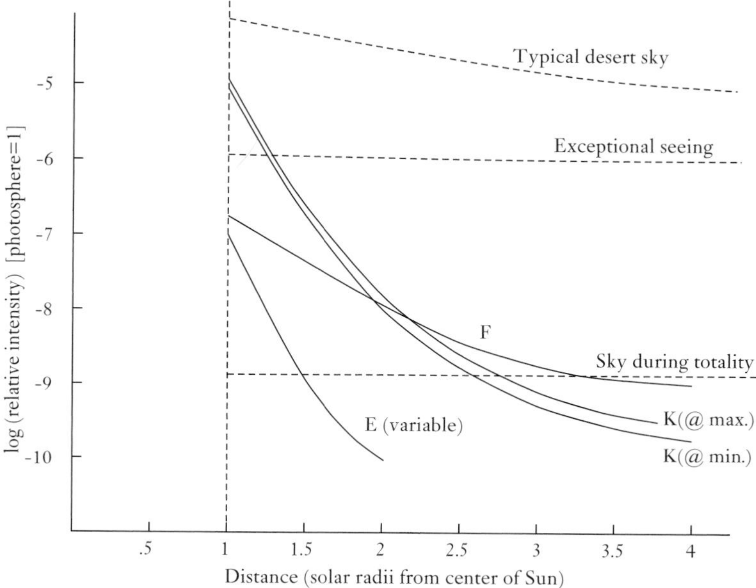

FIGURE 5.13. The main components of the solar corona. Note that the vertical scale is logarithmic (powers of 10) and is relative to the brightness of the solar photosphere. The corona is generally below -6, or one millionth, on this scale.

the dust particles are moving relatively slowly, the light reflected from them is not very much shifted in wavelength relative to the incoming light, and the dark lines remain discernible in the scattered light.

The F-corona is not coronal light, although it happens to be about the same brightness, and it also falls off in intensity as one moves away from the disk of the Sun.

E-corona The emission-line (hence "E-") corona is the only component in this list that is actually due to emission of light from the hot coronal gas. It is due to the excitation of electrons in atoms of the corona, followed by atomic transitions of those electrons back down to lower energy levels, with the consequent emission of light. The emission-line component of the corona is hard to see in broad wavelength recordings of the coronal light – such as photographic plates – but it can be detected clearly if a narrow slice of the spectrum is selected for viewing. Within this small range of wavelengths – which might be anywhere from the infrared through the visible and ultraviolet into the soft x-ray region of the spectrum – the E-corona is relatively bright, and can be recorded for scientific study.

T-corona This newly recognized component of the corona is now detectable because of improvements in infrared viewing techniques. The interplanetary dust that causes the F-corona and the zodiacal light is warmed by the Sun and therefore emits radiation due to its temperature. This thermal (hence "T-") emission adds to the continuum of the spectrum in the infrared.

Only the E-corona would show obvious evidence that the corona is hot, and it is quite faint except for the "coronium" emission lines, which were seen in the 19th century, but which could not be identified at that time. The F- and K-coronas were dominant in photographic images and in low-resolution spectrographs, and they are due to photospheric light. So it is understandable that the coronal spectrum looked photospheric, except for the puzzling emission lines.

Using x-rays from the corona

Although the first true coronal emission that was detected was visible light, this was an historical accident rather than

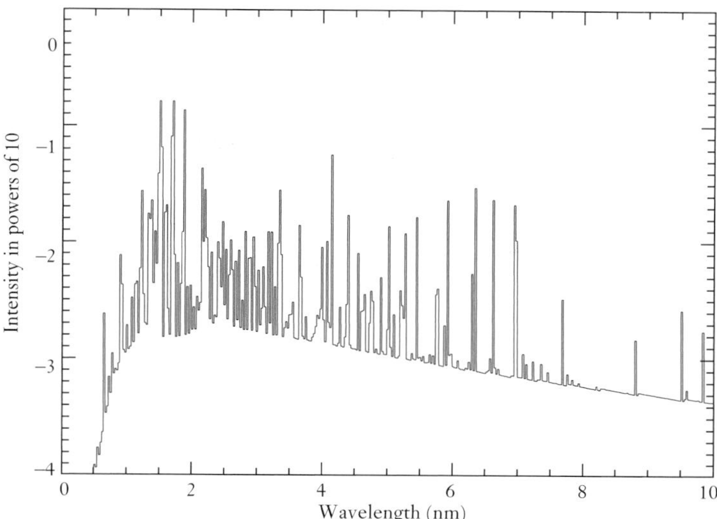

FIGURE 5.14. The spectrum of the coronal plasma at 3 million degrees. The light emitted is about 100 times shorter in wavelength than visible light.

something fundamental about the corona itself. In reality, because of the corona's high temperature, the bulk of the radiation it emits occurs at much shorter wavelengths than those of visible light, in the regions called the extreme ultraviolet (EUV) and also shorter wavelengths known as soft x-rays ("soft" because they are much lower energy and have much less penetrating power than, for instance, medical x-rays).

Figure 5.14 shows the spectrum of the corona at x-ray wavelengths, based on a complex computer code that takes into account dozens of different constituent elements in the corona, dozens of atomic energy levels in each element, and thousands of interactions among all of them, all as a function of temperature. Note that the scale is in nm, or nanometers (1 nanometer = 10 Å), and that the typical wavelength emitted is about 2 nm. For visible light, the typical wavelength would be about 500 nm.

Different parts of the corona are at different temperatures, ranging from 1,000,000 to 10,000,000 degrees, and the spectrum varies with temperature. This spectrum is for a piece of the hot corona in active regions, at a temperature of 3,000,000 K (about 5,000,000°F).

Such calculations show us what the emission is from the hot coronal gas, but not what it *does*. More is needed if we want to know how the gas is heated and why it behaves so dynamically. We need a way to produce an image of the x-ray emission. A telescope that is specially designed to focus x-rays is needed for these wavelengths. Several such instruments, such as the x-ray Telescope (XRT) and the EUV Imaging Spectrometer (EIS) on the Hinode satellite, will be discussed in the next chapter.

SUGGESTIONS FOR FURTHER READING

Golub, Leon and Pasachoff, Jay M., *The Solar Corona, 2nd ed.* (Cambridge University Press, 2010).

Guillermier, Pierre and Koutchmy, Serge, *Total Eclipses: Science, Observations, Myths and Legends* (Chichester: Praxis and Springer, 1999); translated from *Eclipses Totales: Histoire, Découvertes, Observations* (Paris: Masson, 1998).

Levy, David H., *David Levy's Guide to Eclipses, Transits, and Occultations* (Cambridge: Cambridge University Press, 2010).

Littmann, Mark, Willcox, Ken, and Espenak, Fred, *Totality – Eclipses of the Sun*, 3rd edition (Oxford University Press, 2009).

Pasachoff, Jay M., *Peterson Field Guide to the Stars and Planets*, 4th ed. (Houghton Mifflin Harcourt, Boston, 2000, 2012 printing).

Pasachoff, Jay M., and Covington, Michael, *The Cambridge Eclipse Photography Guide* (Cambridge University Press, 1993).

Sheehan, William, and Westfall, John, *Celestial Shadows: Eclipses, Transits, and Occultations* (New York: Springer, 2014).

6

Space Missions

Putting observing instruments into space is difficult, expensive, nerve-wracking, and often frustrating. Launch opportunities are limited, the risk of failure is high, and it usually requires many years for a project to reach completion. Why then would anyone want to do science this way?

The problem is that the Earth has an atmosphere. This is good news for humans, of course. Life as we know it would not be possible without an oxygen-rich atmosphere, which also acts as a shield against harmful solar radiation and energetic particles known as cosmic rays (which are also deflected up toward the poles by the Earth's magnetic field – another type of shield). Our atmosphere prevents most meteors from hitting the ground – except for an occasional larger one – and it provides a greenhouse warming that brings global temperatures up to habitable levels, as we will discuss in the next chapter.

Most of these benefits to life are bad news for astronomers. From ground level we do not have direct access to cosmic rays, so it is difficult to study their origin; we do not have direct access to the bits of dust and rock floating through the solar system, to help determine the history of formation of the planets; we cannot, from the ground, directly measure the solar energetic particles to study the origin and effects of solar storms. Merely trying to look at astronomical objects is difficult: the atmosphere is murky and turbulent, it is full of dust and other obscuring materials, and scattered light from human habitation prohibits observations from most populated areas. The best views of these objects are therefore obtained by putting the telescopes above the atmosphere, as was done with the Hubble Space Telescope.

Our atmosphere is also quite opaque to most wavelengths of light, as we showed in Figure 4.1. It is only the very narrow slice of wavelengths in what we call the "visible" that penetrates the atmosphere fairly well. But astronomical objects emit energy at wavelengths from the radio through the infrared, visible, ultraviolet, x-rays, and even gamma rays. With our extremely limited sensory equipment and sitting at the bottom of our murky ocean of atmosphere we are literally blind to most of what is happening out there.

Another major reason to go above our atmosphere is to decrease the background noise level. An example may help to explain why this matters. If you are at a rock concert, with the performers using 10,000 watt amplifiers, you will not be able to hear a person next to you speaking or even shouting in your ear. On the other hand, if you are in the reading room of a library, someone whispering across the room may seem loud enough to be disturbing. The important quantity in these comparisons is the "signal-to-noise" ratio, the intensity of the desired information (signal) to the undesired information (noise). If the noise level is reduced, we can more easily detect a weak signal.

Our atmosphere is never completely dark, even at night. Sunlight and city lights scatter in air, so that there is always some amount of light coming down to us from the sky above. This can be seen in a long-exposure photograph, showing that there is enough illumination at night to take a picture, if the shutter is left open long enough. But astronomers looking at faint objects do need to take long exposures, sometimes several hours long, so that the sky brightness becomes a factor limiting the faintness of the targets that can be seen.

Solar astronomers trying to record the corona encounter this problem. The corona, even near the Sun, is faint and special instruments are needed to detect it, as we described in the last chapter. The corona quickly becomes even fainter as we look further away from the Sun, and the background sky is very bright since, by definition, we are observing in the daytime. If we want to study the corona as it leaves the Sun, or as hot plasma is ejected from the Sun toward the Earth, we need to have a lower sky background, and this is only obtained in space. A nice example of the advantages gained by placing a coronagraph in space is provided by the view of the extended corona recorded by the LASCO and EIT instruments on SOHO, shown in Figure 6.1.

There are also reasons to carry out observations in the infrared, at wavelengths longer than the visible. First, a large fraction of the energy from the Sun and stars is emitted at wavelengths longer than the visible. In addition, there are features on the Sun and other astronomical objects (such as star formation regions) that, because of their temperature, are best seen in the infrared. But our atmosphere is at a finite and relatively warm temperature, which means that it is a source of infrared radiation itself. If our eyes could see these wavelengths, the night sky would not look dark at all, but would be glowing like the burner on your stovetop. So again, an entire slice of the spectrum becomes available by putting your telescope above the atmosphere, or at least very high up on a mountaintop or using a large balloon in order to be above as much of the atmosphere as possible. Note that the telescope mirror itself must be cooled, or it glows in the infrared as well!

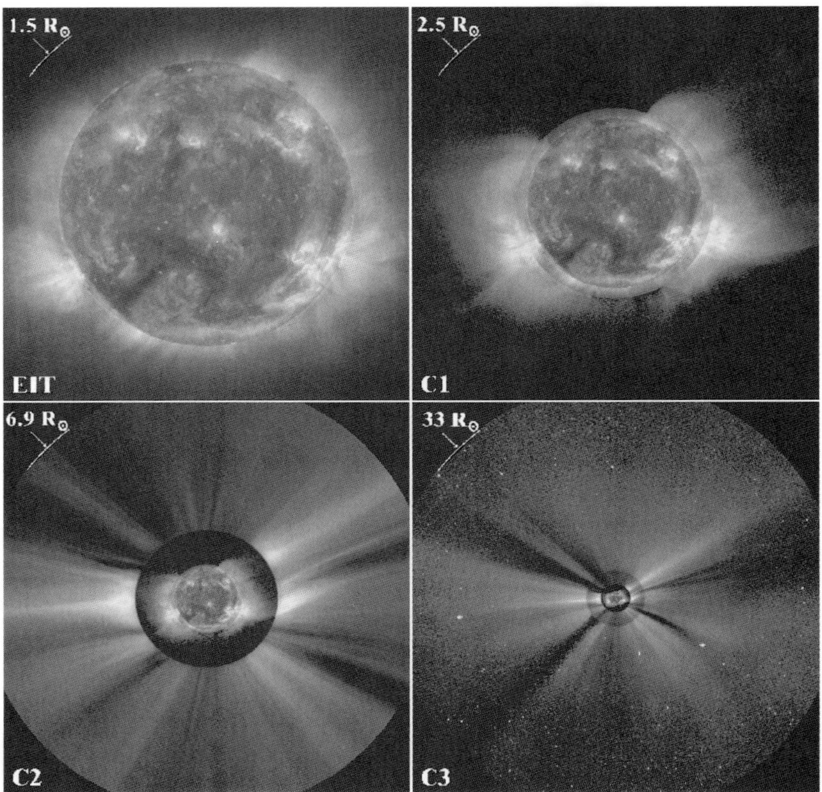

FIGURE 6.1. Combined LASCO and EIT images showing the extended corona.

THE VIEW FROM SPACE

Solar studies from space have been concentrated mainly in two areas: direct measurements of particles coming from the Sun, and observations of the invisible wavelengths of light in the ultraviolet (UV), soft x-ray and the very faint visible parts of the spectrum which are undetectable from the ground. Historically, the particle measurements arose out of the many years that had been spent studying "cosmic rays," and the invisible wavelength measurements came about mainly because of the Sun's influence on radio communications.

The radio studies commenced with the early "wireless" transmissions of Marconi in 1901. Signals were sent succesfully from England to Canada, even though the curvature of the Earth prevents a direct line-of-sight view between such widely separated locations. (It is not clear why he tried this in the first place, given that it should not have worked.) The explanation for this long-range transmission, arrived at independently by an American engineer, Arthur Kennelly, and the British physicist Oliver Heaviside, is that a layer of the upper atmosphere is highly ionized – that is, it contains a large number of mobile electrons, giving it a high electrical conductivity – so that the atmosphere acts as a conducting waveguide that reflects (actually scatters) the radio signal back down toward the ground, a mode now called "Skywave" or "skip." The reflecting layer was later given the name "ionosphere." The obvious next question is, what is causing this ionization of the upper atmosphere? The answer came largely through the work of Edward Hulbert, head of the Heat and Light Division of the Naval Research Laboratory in Washington, D.C. Hulbert established that there were day/night and seasonal variations in the ionosphere, and coupled this with the disruptions in radio communications following solar flares, to suggest that the Sun was the source of the ionosphere. In particular, he suggested in 1938 that x-rays from the Sun were the main contributor to the ionization, the basic idea being that the incoming x-rays are absorbed by the first traces of gas high in the atmosphere and transfer their energy to these atoms, resulting in an electron being "kicked out" of the atom every time an x-ray is absorbed.

Testing this theory required direct, or *in situ*, measurements. Unfortunately, the ionosphere is formed at altitudes between 80 and 225 km (50 to 140 miles), far higher than could be reached by any available methods at the time, such as aircraft or balloons.

But in 1945 the U.S., and later the Russian, armies took over control of the German V-2 missile factory near Penemunde, and shipped trainloads of rocket parts (and rocket scientists) back to their respective home countries. In the U.S., a rocket test site was set up near White Sands, New Mexico, and an offer was made to NRL (which had presciently just created a "Rocket Sonde Research" branch – "sonde" meaning a device for testing physical conditions in remote locations) to use the V-2s for scientific studies. The early flights clarified some of the difficulties of rocket research, such as recovering instruments that had been turned into mangled scraps of metal at the bottom of a large impact crater when the rocket returned to Earth at many times the speed of sound.

By 1952, Herb Friedman of NRL had shown that solar x-rays are indeed responsible for the formation of the so-called "E-region" in the upper atmosphere. Friedman also branched out into night-time rocketry and helped establish the field of x-ray astronomy, although it was a group in Cambridge, MA, led by Bruno Rossi and Riccardo Giacconi, that actually detected the first non-solar x-ray source. In 1957, space research in the United States was boosted by the shock of learning that the Soviet Union had put a satellite ("Sputnik") into orbit; the West did not, at the time, think that the Soviets had such an advanced missile capability. Within a year, the U.S. created a civilian space agency, NASA, to promote space research, with a large degree of independence from the military's own space programs.

Wind from the Sun

It is well known these days that a comet's tail always points away from the Sun, no matter where in its orbit the comet happens to be. The long tail seems to act like a wind sock at a weather station, billowing out to show the direction of the wind. In the early

1950s the prevailing theory to explain this effect was that "radiation pressure" from the solar light was the cause. Einstein had shown at the start of the 20th century that photons of light carry momentum, just as do high-speed particles. If there is enough light radiating out from the Sun, then it will push on the comet tail, causing it to point away from the Sun. This process has, in fact, been proposed as the propulsion method for deep space probes, using a "solar sail."

However, the German astrophysicist Ludwig Biermann calculated that the radiation pressure from the Sun was far too small to produce the observed effect. He noted that there is at all times a persistent weak aurora at high latitudes over the Earth, and there are also persistent small variations detected in the Earth's magnetic field. The idea was developing that these effects were due to a continual bombardment of the Earth by charged particles streaming out from the Sun. Biermann asked whether such a particle stream could be responsible for the comet-tail effect, and he calculated that a high-speed outflow, at a speed of roughly 300–600 miles per second, and a density of about 1,000–10,000 particles per cubic inch, would do the job. The outflow would represent a substantial portion of the coronal mass, which would have to be replenished from the photosphere daily, and the Sun would be constantly losing mass. However, the rate is small enough that the Sun would lose only a tiny fraction of its mass over its 10 billion year lifetime.

This idea was not immediately accepted, because there was no obvious theoretical explanation for such an outflow. This was provided in 1957 by the work of Sydney Chapman, who showed that a hot corona will extend far out into space, and by Gene Parker, who showed that a hot corona will expand outward continually. Chapman showed that the coronal plasma, because of the high mobility of its electrons, has the ability to conduct heat readily over long distances. If the part of the corona near the Sun is hot, then he showed that the temperature will stay high far out into space, and the corona itself will extend very far away from the solar surface. Parker then calculated that such an extended corona will do more than merely reach far out into space. He predicted that the solar corona will actually expand and flow outward, much as Biermann postulated, and he named this outflow the "solar wind."

Parker's theory created a storm of controversy (he was more than once introduced at lectures with the phrase "He who sows the wind shall reap the storm.") and it was not until instruments were put into space that the high-speed particles were detected and the theory was spectacularly confirmed. Many of the earliest solar observations from space were carried out by Soviet groups, such as that of S. L. Mandel'shtam at Moscow's Lebedev Physical Institute and Konstantin Gringauz at the Radio Engineering Institute. The results ranged from measurements of the solar UV and X-radiation by Sputnik-2 to detection of the solar wind by Lunik-2 in 1959.

However, these early results were rudimentary and not widely publicized in the West. It required a more elaborate instrument, flown by Bruno Rossi's group on Explorer-10, and then a much longer and more complete set of measurements flown on NASA's first successflul planetary mission, Mariner-2, by Conway Snyder and Marcia Neugebauer in 1962, before the existence of the solar wind was fully acknowledged.

The difference it makes

The photosphere, or visible surface of the Sun, looks fairly blank and uniform to the naked eye, although those with very good eyesight can occasionally see particularly large sunspots (viewed at sunset, or with a dark filter to protect your eyes). Through a small telescope sunspots are easily visible, and best viewed by projecting the image onto a screen. But there is far more happening than is seen in visible light: Figure 6.2 shows how much difference it makes to have the x-ray wavelengths available.

The upper image in this figure shows the solar surface as it might appear from a ground-based telescope (actually, this photo was taken in the near ultraviolet by the TRACE satellite, but it shows pretty much the same height in the solar atmosphere that you see in white light, and it has the advantage of being simultaneous, coaligned and at the same magnification as the coronal image). The sunspots are easily visible, and in fact there is a group of sunspots in the lower right-hand portion of the image. With a good solar telescope, placed at an exceptionally clear observing site, sunspots show a characteristic detailed structure, as seen in the top panel of Figure 6.2. The center of the spot, known as the "umbra," is quite dark, and it is surrounded by an iris-like "penumbra." The rest of the surface is covered with a fine mottling, called *granulation*, which is caused by convective motions (like those of water boiling in a pan) of the surface. In this image one can also see something not usually found in white-light images: there are patches of brightness near the sunspots, and these are found to correlate very closely with the corona above the surface. Both the sunspots and the bright patches are caused by strong concentrations of magnetic fields breaking through the solar surface from inside the Sun.

The second photo shows the hot corona above the solar surface, and it is fairly obvious that one would not have suspected, from looking at the photosphere, that the atmosphere above it is so rich in structure. The coronal image is noticeably different, showing large loop-like structures, and long thin rays radiating out of the frame. All of these features are at temperatures of several million degrees and the material filling these structures is a *plasma*, or ionized gas.

The x-ray light that we see coming from these regions is emitted mainly by electrons making a transition from one energy level in an atom down to another lower level. In this case, the

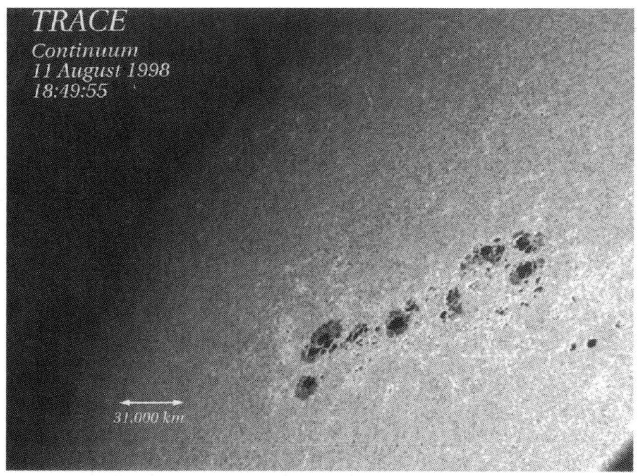

FIGURE 6.2. Comparison of the solar surface (top photo), which shows sunspots and granulation, with an image of the hot corona (bottom photo), which shows the solar atmosphere above the surface. Sunspots are places where strong magnetic fields from inside the Sun break through, and the hot corona traces out the magnetic field above the surface.

transitions are from energy levels not usually seen on Earth. For instance, the lower panel of Figure 6.2 shows emission from a spectral line of Fe IX, eight-times ionized iron, which is only produced in plasmas that are at a temperature of about one million degrees (kelvins). (Note that the spectrum of neutral, or zero-times ionized iron is labelled Fe I, so the two ways of counting give values differing by one.) One does not normally encounter temperatures of millions of degrees on Earth, except in specially-produced lab experiments, and in thermonuclear explosions.

ARCHITECTURE OF A SPACE MISSION

If you are an amateur astronomer, it may be difficult to picture the problems associated with putting observing equipment in space. On the ground your basic procedure is: set up the telescope, point it at the desired object, put your eye to the eyepiece and observe. In space, things are different:

1. First, there is no place to set up the telescope. It is floating weightless in a vacuum.
2. Next, you cannot use your hand to push the tube into position since you are not there.
3. Likewise, you cannot put your eye to an eyepiece, since it is several hundred miles, or further, above your head, moving at tens of thousands of miles per hour.

These are only the most obvious difficulties. Your instrument is also reaching temperatures of several hundred degrees on the side facing the Sun and is approaching Absolute Zero on the anti-sunward side. In order to get into space, it had to survive tremendous acceleration and vibrational forces from the rocket motors, which would severely damage an ordinary "terrestrial" instrument. If you have a computer onboard, its processor is probably overheating, because the cooling fan is useless in a vacuum, and it will fail soon anyway because cosmic ray damage will destroy the electronic circuitry.

Sending equipment into space therefore involves difficulties that we do not normally encounter in ground-based operations. A typical space observatory must have these components:

- The scientific instruments one wishes to use.
- A "spacecraft," which is a platform that typically provides power, telemetry and pointing capabilities to the instruments.
- A launch vehicle to get these items into space.

The scientific instrument is thus part of a larger package that gets the instrument into space, points it at the correct target, provides power to run the cameras and related electronics, and sends the data down to the ground via radio contact.

Sounding rockets

Most of the steps involved in putting a satellite into orbit are also found in the quicker and simpler sounding rocket launch, so we will first discuss that type of mission before going on to satellites. Figure 6.3 shows a typical sounding rocket payload, as an example of a launch configuration. Sounding rockets are sent up above the atmosphere for brief periods of time – only a few minutes – but have the same set of difficulties to overcome, since they must supply their own power, pointing, radio contact with the ground (telemetry), plus they should also have a recovery system if one wants to refly the instrumentation. Sounding rocket flights are brief because the rocket does not attain the speeds needed for orbiting the Earth, but they are far less expensive than orbital experiments and so have easier access. The trade-off is that they are only tossed up into space briefly and then fall back to Earth.

In a sounding rocket flight we go through the same steps needed to put a usable satellite telescope into orbit, except that

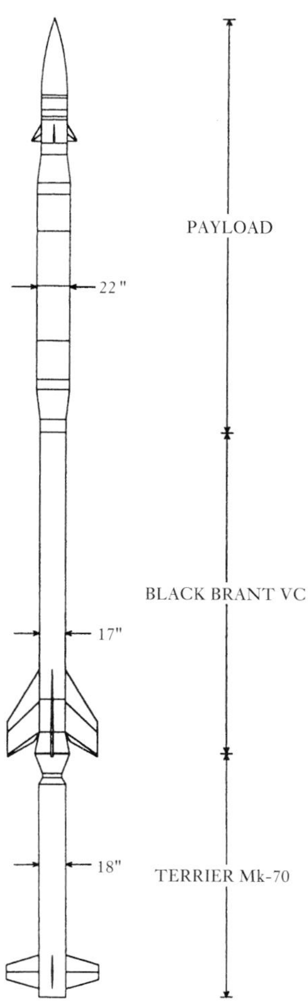

FIGURE 6.3. Drawing of a typical sounding rocket payload used for solar studies. The two stages of rocket motors are at the bottom; the telescope (payload) is in the middle, and the power, pointing, telemetry and recovery sections are at the top.

everything happens in a few minutes, instead of days or weeks. Because a sounding rocket flight only lasts a few minutes, there is no time for a leisurely checkout and turn-on. The design philosophy changes accordingly. For example, the skin of the rocket becomes tremendously hot during ascent due to air friction, and then the side of the instrument facing the Sun becomes very hot in vacuum because there is no air to take the heat away. In a satellite, the instrument is designed to reach an equilibrium condition, a balance between heating and cooling which may take hours to achieve. In a sounding rocket, the instrument is instead designed so that the heat takes as long as possible to get into the instrument; the observations are over before the telescope knows what hit it.

As an example of the steps involved in a countdown, launch and recovery, Table 6.1 gives a listing of the major events involved in the launch and recovery of one of our telescopes.

The pre-launch activities include checkout of all systems, and a "vertical" test, in which it is verified that signals can be sent to and from the rocket. At T–5 minutes, the umbilical lines to the payload are disconnected and the rocket goes onto internal battery power for the remainder of the flight. It is interesting to note that the famous countdown to zero that you see any time a rocket is launched is not merely used for dramatic effect. As a participant in the launch, you actually need this method in order to focus your concentration on the crucial set of steps that must be carried out in strict accordance with the timeline checklist. It is quite useful to have someone counting for you, so that you can carry out your part of the checklist without having to worry about looking at the clock. In addition, to paraphrase Samuel Johnson[1], the approach to zero serves to wonderfully concentrate the mind.

Note that in the case of this sounding rocket, the first stage burns for only about 6 seconds and that the vehicle accelerates to nearly twice the speed of sound in that time. Clearly there are fearsome g-forces involved, but an even more difficult problem

[1] "Depend upon it, sir, when a man knows he is to be hanged in a fortnight, it concentrates his mind wonderfully."

Table 6.1 Flight Day Checkout Procedure & Event Timeline

Time	Event	Mach Number
T–240 min	Begin experiment evacuation	
T–90	All personnel at stations	
T–80	Tower function checkout	
T–65	Close vacuum valve	
T–25	Go for launch	
T–15	Verify data recorders functional	
T–5	Experiment and telemetry on internal power	
T–4	Verify telemetry contact	
T–2	Programmed launch hold	
T–0	Launch	0
T+0.6 sec	Rail release	0.1
T+6.2	First stage burnout	1.7
T+12	Second stage ignition	1.4
T+18	S-19 decouple	2.1
T+30	Verify experiment vacuum valve open	
T+44.4	Second stage burnout	6.4
T+50	Experiment cameras on	
T+61	Payload despin	6.7
T+64	Nose Tip eject	6.8
T+66	S-19 gyro caged	6.8
T+68	Attitude Control System on	6.8
T+89	Verify experiment detectors on	
T+95	Sun acquired	
T+96	Verify pointing	
T+98	Start of data-taking	
T+268	Apogee	0.4
T+457	Loss of pointing; end data-taking	5.6
T+506	Mach 1	1
T+510	Parachute open	0.03

is "vibration." Equipment will be literally shaken to pieces during these few seconds if not properly designed. At the very least, precision optics can be thrown out of alignment, electronic circuit boards can crack, or a loose screw left in the payload can come crashing through a vital piece of hardware. (In one case, an experimenter discovered after recovery that a bug had crawled into the payload; after the flight, it was found flattened against the rear bulkhead, as if someone had stepped on it.)

During first-stage burn the vehicle is spun up for stability. When the first stage burns out it separates from the rest of the rocket and drops away. The second stage then ignites and brings the payload up to to a speed of about Mach 7, then it too burns out and separates from the payload. Once up above the atmosphere, about a minute after launch, the payload is in free-fall, following a ballistic trajectory upward to some maximum height, then falling back down into the atmosphere. The payload is then de-spun and a door opened so that the telescope can make observations, gas jets swing it around to point at the Sun, and the attitude control system (ACS) takes over for precise solar pointing. Fortunately, the Sun is the brightest object in the sky and it is quite circular. A fairly simple sensor can therefore find the Sun and point to the center of the disk with good reliability and high precision.

All of this happens in under two minutes! The total amount of time "at altitude" for data-taking is about 6 minutes (457 – 98 = 359 seconds). By coincidence, this is about the duration of the longest total eclipse, so the experience for both is similar: it is all over almost before you have had time to think. In fact, one of the most important lessons that a rocket experimenter or an eclipse observer must learn is that there will not be any time to think during the observation period. Every move must be planned out ahead of time, and responses to all of the possible emergencies that could arise – for instance, finding that the pointing system has not pointed correctly at the Sun – must be planned and practiced in advance.

The *most* important lesson that a space experimenter must learn is that once the instrument on the ground is closed up for launch you will not be able to touch it again. This means that you must make absolutely certain that it will work as expected without further adjustments or tinkering (except for a very few things that you can do by remote telemetry command, such as changing the pointing direction of the telescope, or contacting the instrument control computer to alter the planned sequence of observations). In the case of a sounding rocket, the instrument only needs to work for five minutes, but you had better be sure that it works for the right five minutes. For a satellite, it must work for months or years, under very harsh conditions. While these considerations may seem obvious, a mindset very different from that of the laboratory worker is required in order to put these warnings into practice from the outset, at the initiation of the design stage. This explains why there is a special job title called "aerospace engineer."

A technique for imaging the corona in soft x-ray light, known as multilayer coating, was originally tried out on sounding rockets, and has now been used in satellites. In this method, a series of extremely thin and very precise layers are deposited onto a mirror to make it reflective to x-rays. The advantage of a sounding rocket is that it is a relatively fast and inexpensive way to get instruments into space, and to demonstrate that a new technique will work, as was done for this new imaging method. The disadvantage of rockets is that you only get about five minutes of observing time. However, a few minutes' duration is infinitely better than no observing time at all.

Satellites

Satellites are an order of magnitude more expensive than sounding rockets, but they can carry out observations for many months or even years. One can therefore plan a far more extensive program of scientific studies with a satellite, including coordinated observations with other satellites and ground-based observing facilities. Figure 6.4 shows the TRACE satellite, which used the multilayer technique for imaging various wavelengths of coronal radiation. It was launched on April 2, 1998,

FIGURE 6.4. The TRACE observatory, shown mounted in the Pegasus spacecraft just prior to launch in 1998. NASA systems engineer Joe Burt smiles approvingly.

and was shut down on June 21, 2010, after the launch and commisioning of the Solar Dynamics Observatory, discussed below. The orbit in this case was "polar sun-synchronous," which means that the satellite moves in a north-south direction, from one pole of the Earth to the other, and that the plane of the orbit faces the Sun at all times.

This type of orbit has the advantage that it is possible to have continuous, 24-hour per day, viewing of the Sun. A disadvantage is that the satellite passes through the Earth's radiation belts (see Chapter 8), which produces noise in the data and can harm the electronics onboard; but the advantage of uninterrupted viewing of dynamic changes and flare events in the corona greatly outweighs the negatives.

There is also a subtlety to the orbit: if the satellite is injected into a straight up-and-down (90 degree) orbit, then the plane of this orbit will remain fixed in space. As the Earth moves around the Sun during the year, the satellite will no longer point at the Sun. It will instead remain pointing in a constant direction relative to the fixed stars, while the Sun will appear to move completely around the Earth, relative to the plane of the orbit. The satellite is therefore launched into an orbit that is slightly more inclined to the poles (about 98 degrees) and the bulge of the Earth's equator causes a very slow drift in the plane of the orbit (known as a "precession") so that the orbit stays locked onto the sunward direction.

FUTURE MISSION PLANS: THE SUN-EARTH CONNECTION

NASA's Office of Space Science (http://www.hq.nasa.gov/office/oss/) has developed four principal science themes, including the Heliophysics (formerly Sun-Earth Connection) program, which includes the scientific disciplines of solar and heliospheric physics, magnetospheric physics, and aeronomy (the study of the upper atmospheres of the Earth and other planets). The main focus of this program is to study the physical processes that link the Earth to the Sun, but it now also includes the other planets in the solar system as well as the study of the outer reaches of the Sun's influence into interstellar space.

The Heliospherics program, which is connected with a major NASA initiative named "Living with a Star," is divided into four "Quests" for mission planning purposes:

Quest 1. How and why does the Sun vary?
Quest 2. How do the Earth and planets respond to solar variability?
Quest 3. How do the Sun and Galaxy interact?
Quest 4. How does the Sun affect life and society?

The process of developing a plan for actual space missions includes input from the scientific community, the formation of science working groups to develop detailed descriptions of proposed missions, additional higher-level scientific panels to narrow down the list to a manageable number, discussions within NASA and with Congress and OMB to set priorities, leading eventually to an Announcement of Opportunity issued by NASA. The latter is an invitation to scientists to propose scientific investigations for a particular mission, and these proposals are then peer-reviewed by panels of scientific and technical reviewers to determine the winning proposals.

All of this clearly takes a long time, and it can be many years from the time a scientist thinks of a study that ought to be carried out until an instrument is flown to acquire the needed data. NASA has recently completed a review of their Heliophysics plans, called the 2013 Heliophysics Roadmap. The plan may be viewed at http://sec.gsfc.nasa.gov/sec_roadmap .htm.

The Living with a Star initiative includes the first of its four major missions, the Solar Dynamics Observatory, and the Radiation Belt Storm Probes program launched August 30, 2012, to study the Van Allen radiation belts circling the Earth; the mission has been renamed Van Allen Probes to honor the discoverer of the radiation belts. The LWS program is focussed on understanding those aspects of the Sun and of our terrestrial space environment that most directly affect life on Earth. The LWS program has a Web site at lws.gsfc.nasa.gov/, the SDO Mission Web site URL is sdo.gsfc.nasa.gov/, and the Van Allen Probes Web site may be found at www.nasa.gov/ rbsp/.

The Heliophysics program involves activities with impact far beyond NASA's own programs. The range of implications from this program may be seen in Figure 6.5.

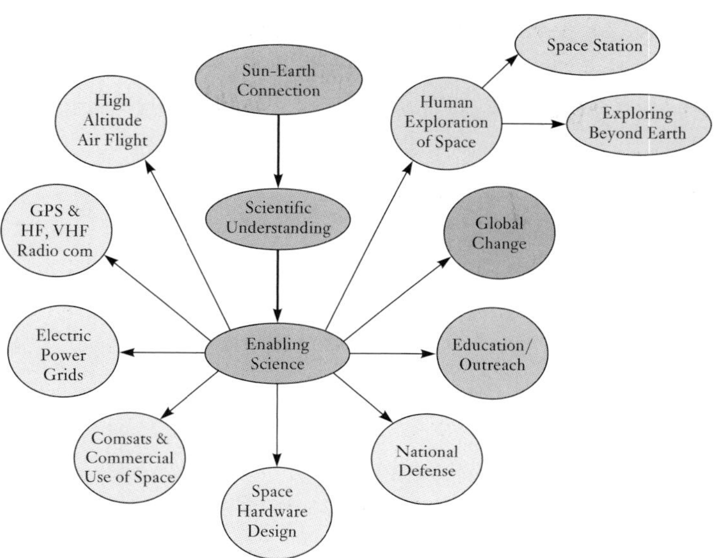

FIGURE 6.5. The connection between the Heliophysics (formerly Sun-Earth Connection) program and areas as diverse as electric power distribution to manned space flight are shown in this chart.

Current and near-future mission plans

At any given moment, there are active space missions whose satellites are operational and sending data back to Earth, as well as missions under construction and being readied for launch. In addition, there are missions in the planning stages, either approved and in the design phase, or under consideration and not yet started. In this section we describe some of the solar missions currently in operation, some nearing completion, and a few mission ideas for the more distant future.

RHESSI

The Reuven Ramati High Energy Solar Spectroscopic Imager (RHESSI) is designed to study the particle acceleration and energy release in solar flares. RHESSI does this by

FIGURE 6.6. The fully integrated spacecraft and science instrument for NASA's Interface Region Imaging Spectrograph (IRIS) mission is seen in a clean room at Lockheed Martin Space Systems. The solar arrays are deployed in the configuration they assumed in orbit (2013 launch). (See p. 209.)

concentrating on the energetic electrons and protons produced in flares, through observations of the x-rays and gamma rays that these particles produce. The range of energies detected by RHESSI is from soft x-rays, down to 3 thousand electron volts (eV), up to gamma rays at 20 million eV.

The RHESSI mission consists of a single spin-stabilized spacecraft – that is, pointing at the Sun is maintained by a gyroscope-like effect, from spinning of the entire instrument around the axis pointing to the Sun – in a low-altitude orbit inclined 38 degrees to the Earth's equator. The only instrument onboard is an imaging spectrometer – one that provides both an image and a spectrum at the same time – with the ability to obtain color movies of solar flares in x-rays and gamma rays. The word color here is used in the broad sense of being able to separate different wavelengths in the image.

RHESSI has the finest spatial and spectral resolutions of any hard x-ray or gamma ray instrument ever flown in space. The mission is described in more detail on the RHESSI Web pages (http://hessi.ssl.berkeley.edu/ and http://hesperia.gsfc.nasa.gov/hessi/index.html). RHESSI was launched into a polar, sun-synchronous orbit on February 5, 2002. Since its launch RHESSI has observed over 8,000 solar activity events.

IMAGE

Imager for Magnetopause-to-Aurora Global Exploration (IMAGE – Fig. 6.7) is a MIDEX (mid-size Explorer) mission, selected by NASA in 1996 to study the global response of the Earth's magnetosphere to changes in the solar wind. IMAGE is the first satellite mission dedicated to imaging the Earth's magnetosphere, the region of space controlled by the Earth's magnetic field and containing extremely tenuous plasmas of both solar and terrestrial origin. IMAGE uses neutral atom, ultraviolet, and radio imaging techniques to:

Identify the dominant mechanisms for injecting plasma into the magnetosphere on substorm and magnetic storm time scales;

Determine the directly driven response of the magnetosphere to solar wind changes; and,

Discover how and where magnetospheric plasmas are energized, transported, and subsequently lost during substorms and magnetic storms.

In order to fulfill its science goals, IMAGE utilizes neutral atom, ultraviolet, and radio imaging techniques. A suite of three neutral atom imagers (NAI) provides energy- and composition-resolved images at energies from 10 eV to 200 keV with a time resolution of 300 seconds. Two ultraviolet imagers, covering wavelength ranges from 120–180 nm (FUV) and 30.4 nm (EUV), provide coverage in the far and extreme ultraviolet. A radio plasma imager (RPI) transmits and receives pulses from 3 kHz to 3 MHz allowing relative motions of the satellite and

FIGURE 6.7. The IMAGE spacecraft during pre-launch ground testing.

plasma to be determined to a resolution of 400 m/s and a time resolution as good as 4 s. The IMAGE satellite was launched March 25, 2000. More information on this mission is available at: http://image.gsfc.nasa.gov/

TIMED

The Thermosphere * Ionosphere * Mesosphere * Energetics and Dynamics (TIMED) Program is the first science mission in the Solar Connections Program as detailed in NASA's Strategic

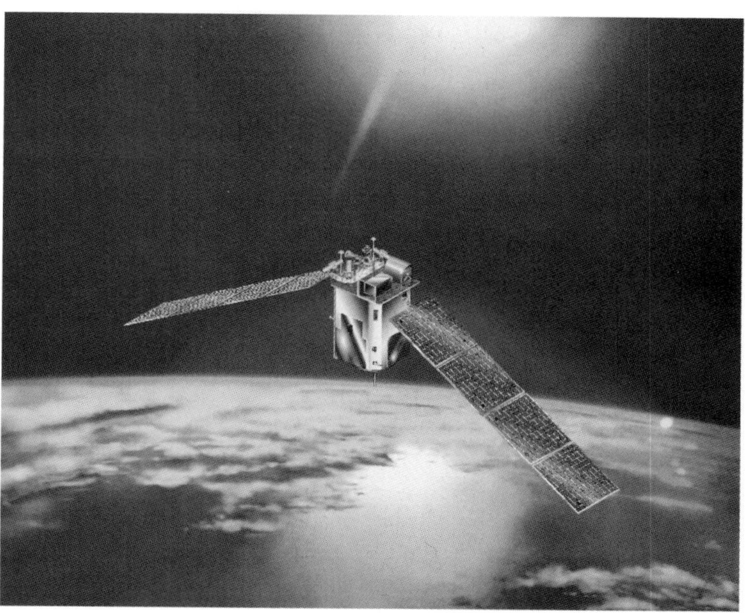

FIGURE 6.8. Artist's concept of the TIMED spacecraft.

Plan. TIMED (Fig. 6.8) explores the Earth's Mesosphere and Lower Thermosphere (60–180 kilometers), the least explored and understood region of our atmosphere. It is known that the global structure of this region can be perturbed during stratospheric warmings and solar-terrestrial events, but the overall structure and dynamics responses of these effects are not understood. Advances in remote sensing technology employed by TIMED instrumentation allows us to explore this region on a global basis from space.

The primary objective of the TIMED mission is to investigate and understand the energetics of the Mesosphere and Lower Thermosphere/Ionosphere (MLTI) region of the Earth's atmosphere. The MLTI is a transition region in which many important processes are known to change dramatically, yet there are only sketchy and incomplete measurements of the region. This is where temperature variations, which increase dramatically with altitude, are largest, and where atmospheric motions are dominated by poorly understood gravity waves and tides. When the mission is complete, it will have:

Performed the first focused exploration of the MLTI region – a major link in the solar-terrestrial chain.

Traced the flow of energy and momentum from the lower into the upper atmosphere.

Quantified the dramatic influences of the Sun and outer space on the Earth's middle and upper atmosphere, increasing our ability to predict space weather.

Examined the major energy sources, transport processes and energy sinks in the MLTI, and applied this information to define the channels by which energy enters into, travels through, and is lost from the region.

Supplied knowledge of the upper atmospheric circulation pattern and its role in the transport of chemicals important in ozone chemistry.

TIMED was launched 7 December 2001, and has been approved to continue operations through 2014. Additional information about the mission can be found at the NASA web site: http://science.nasa.gov/missions/timed/

STEREO

The Solar TErrestrial RElations Observatory (STEREO) mission (Plate XI) is the third in the line of Solar-Terrestrial Probes (STP) and is a strategic element of the Heliophysics Roadmap. STP is a continuous sequence of flexible, cost-capped missions designed for sustained study of critical aspects of the Sun-Earth system. It is an outgrowth of the highly successful International Solar-Terrestrial Physics (ISTP) program.

Coronal Mass Ejections, or CMEs, are enormous clouds of hot magnetized plasma hurled from the Sun at speeds up to a million miles an hour (see Chapter 8 for details). The primary goal of the STEREO mission is to advance the understanding of the three-dimensional structure of the Sun's corona, especially regarding the origin of CMEs, their evolution in the interplanetary medium, and the dynamic coupling between CMEs and the Earth environment. CMEs are the most energetic eruptions on the Sun, are the primary cause of major geomagnetic storms, and are believed to be responsible for the largest solar energetic particle events. They may also be a critical element in the operation of the solar dynamo because they appear to remove dynamo-generated magnetic flux from the Sun.

The unique aspect of the STEREO mission is that it employs two nearly identical spacecraft sent respectively ahead of and behind the Earth in its orbit, thereby providing a stereoscopic view of events leaving the Sun as they progress toward Earth. STEREO is, therefore, continuing the systematic study of the relationship between processes on the Sun and their consequences for the Earth. The ejections of well-defined clouds of plasma from the corona were discovered in 1973 by instruments on NASA's Skylab/Apollo Telescope Mount. Although studies continued with the NASA Solar Maximum Mission and the ESA-NASA Solar and Heliospheric Observatory (SOHO) in the ISTP program, these investigations were limited to vantage points that best showed those CMEs directed perpendicular to the line of sight, and therefore that missed the Earth. However, with two spacecraft sent in opposite directions away from the Sun-Earth line, the STEREO mission finally allows unambiguous observations of those CMEs that directly impact the Earth. STEREO has also for the first time provided a stereoscopic view of the three-dimensional corona and the interplanetary medium and thereby advanced the Sun-Earth Connection understanding of the heliosphere begun by the ISTP program.

The STEREO mission launched in October 2006. The latest views of the Sun – now of the side not visible from Earth – can

be found at:
http://www.nasa.gov/mission_pages/stereo/main/index.html

Hinode

The Hinode (Japanese for "Sunrise"; formerly Solar-B) mission is a follow-on to the highly successful Japan/US/UK Yohkoh (Solar-A) collaboration. The observatory (Plate XII) consists of a coordinated set of optical, EUV, and x-ray instruments that apply a systems approach to the interaction between the Sun's magnetic field and its high temperature, ionized atmosphere. The result is an improved understanding of the mechanisms which give rise to solar magnetic variability and how this variability modulates the total solar output creating the driving force behind space weather.

Hinode is, for the first time, providing quantitative measurements of the full vector magnetic field on small enough scales to resolve elemental flux tubes. The field of view and sensitivity allow changes in the magnetic energy to be related to both steady state (coronal heating) and transient changes (flares, coronal mass ejections) in the solar atmosphere. Hinode is an excellent opportunity for highly leveraged US participation in a major mission that is greatly advancing our understanding of the crucial first link in the Sun-Earth connection. Started in 1999, Hinode launched on September 22, 2006, by the Japanese Aerospace Exploration Agency (JAXA), and is approved at this writing through 2016.

The science objectives of the Hinode mission are:

1. Understand the creation and destruction of the Sun's magnetic field: Magnetic fields permeate all space and play an important role in shaping the universe on all scales. The fields are continuously being generated by dynamos in stellar interiors and swept out into space by stellar winds. The solar dynamo

is sufficiently near and operates on a short enough period, 11 years, that it can be studied directly.

2. Study the modulation of the Sun's luminosity: During the last decade observations from space have led to the profound discovery that the total output of energy from the Sun is not constant but varies in phase with the magnetic activity cycle. The amplitude of this variation, over the single cycle measured, was small but only a factor of three to five below the level required for a significant climatic response. Hinode is making the first observations with resolution, wavelength coverage, and time span adequate to determine the mechanism for the magnetic modulation of solar luminosity.

3. Understand the generation of UV and x radiation: The Sun is a powerful and highly variable source of ultraviolet, x-rays and energetic particles, which are known to have major effects on our environment. This high energy radiation must be due to the annihilation of magnetic energy in the Sun's atmosphere, the chromosphere and corona. Due to its broad complement of instruments with high spatial and spectral resolution, Hinode has been able to study processes such as magnetic reconnection and wave dissipation that are believed to be responsible for the conversion of magnetic energy into UV and x radiation.

4. Understand the eruption and expansion of the Sun's atmosphere: The million-degree corona continually expands outward, becoming a supersonic wind that blows past the Earth, buffeting the geomagnetic field and energizing the upper atmosphere. In addition, large parts of the corona are seen to erupt, blasting through the solar wind and causing major magnetic disturbances at Earth. Hinode is providing accurate measurements of magnetic fields, electric currents, and velocity fields that supply the corona, thus revealing the root causes of the Sun's eruptions.

This description, and additional information, can be found at the Hinode home page at the Marshall Space Flight Center: http://hinode.msfc.nasa.gov/.

In order to achieve these scientific objectives, the satellite carries the following instruments:

1. The SOT: a 0.5 m (20-inch) Solar Optical Telescope equipped with a vector magnetograph, narrow band imager and spectrometer to obtain photospheric magnetic and velocity fields at 0.2 arc sec (~150 km) resolution. See Figure 6.10.
2. The XRT: An x-ray telescope to image the coronal plasma in the range 1–20 million degrees at 2 arc sec resolution and large field of view.
3. The EIS: An EUV imaging spectrometer to obtain plasma velocities to an accuracy of 10 km per second along with temperatures and densities in the transition region and corona at 2 arc sec resolution.

The X-Ray Telescope (XRT)

The basic goals of a soft X-Ray Telescope (XRT) for the Hinode mission are to facilitate the study of the dynamics of fine-scale coronal phenomena, such as magnetic reconnection and coronal heating mechanisms, while at the same time recording the large-scale global phenomena, such as coronal mass ejections. In order to meet these objectives, the XRT works closely with the focal plane instruments of the Solar Optical Telescope (SOT) and with the EUV Imaging Spectrometer (EIS).

The U.S. and Japanese Hinode science teams have identified key problem areas to be addressed by this mission. Among those in which the XRT plays a major role are:

1. **Flares & Coronal mass ejections.** How are they triggered, and what is their relation to the numerous small eruptions of active region loops? What is the relationship between

large-scale instabilities and the dynamics of the small-scale magnetic field?

2. **Coronal heating mechanisms.** How do coronal loops brighten? Are wave motions visible, and are they correlated with heating? Do loops heat from their footpoints upward, or from a thin heating thread outward? Do loop-loop interactions contribute to the heating?

3. **Reconnection & coronal dynamics.** *Yohkoh* observations of giant arches, jets, kinked and twisted flux tubes, and microflares imply that reconnection plays a significant role in coronal dynamics. With higher spatial resolution and with improved temperature response, the XRT is helping clarify the role of reconnection in the corona.

4. **Solar flare energetics.** Although Hinode was launched after solar maximum, there were still many flare events seen. The XRT is designed so that it can test the reconnection hypothesis that has emerged from the *Yohkoh* data analysis.

5. **Photosphere/corona coupling.** Can a direct connection be established between events in the photosphere and a coronal response? To what extent is coronal fine structure determined at the photosphere?

In order to address these scientific problems, Hinode has a grazing-incidence X-Ray Telescope (XRT), similar to that flown on *Yohkoh*, but with several improvements:

- The pixel size is improved from 2.46 arcseconds to 1.0 arcseconds;
- The detector resolution is increased from $1K \times 1K$ to $2K \times 2K$; and
- The temperature sensitivity is augmented with a low-temperature ($1-2$ MK) response in addition to an extended high-T response.

In comparison with the *Yohkoh* SXT, the Hinode XRT therefore has six pixels for every SXT pixel, comparable field of

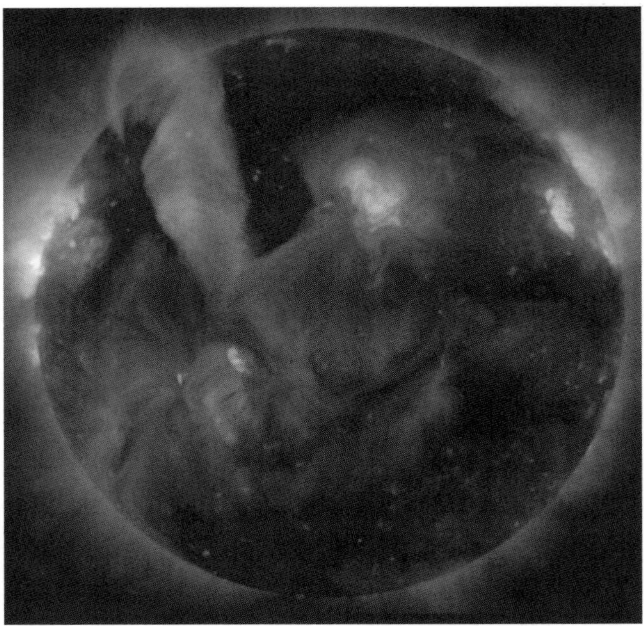

FIGURE 6.9. The high-temperature x-ray corona as seen with the Hinode XRT (x-ray telescope), taken on September 25, 2012.

view, comparable temperature diagnostic accuracy, and a greater range of temperature coverage. An example of the view of the high-temperautre active corona now obtainable with the XRT is shown in Figure 6.9.

The EUV Imaging Spectrograph (EIS)

EIS consists of a multilayer coated single mirror telescope, and a stigmatic imaging spectrometer incorporating a multilayer coated diffraction grating. What this means is that the instrument not only forms a spectrum, but that there is a long narrow slit, along which an image of the Sun is formed, while a spectrum from each point along the slit is formed onto a detector in a direction perpendicular to the slit. One therefore obtains spectra of a slice of the Sun, rather than only of a single small box.

FIGURE 6.10. The superb image quality of the Hinode SOT is evident in this image of Venus crossing the limb of the Sun taken during the transit of June 5, 2012. Note the thin bright arc of Venus's atmosphere visible in the image.

The image of the Sun is produced by the primary mirror which focusses the light onto a slit. On the other side of the slit, there is a curved diffraction grating, which is positioned such that it takes the light coming through the slit and re-images it onto a detector while simultaneously diffracting it according to wavelength.

The Solar Optical Telescope (SOT)

The largest instrument in Hinode is the Solar Optical Telescope (SOT), which has a package of instruments, the Focal Plane Package (FPP), located at the optical telescope's focus.

The optical telescope has a mid-size primary mirror, 0.5 meter diameter, and is of Cassegrain design. This gives a long focal length, so that the image of the Sun at the focal plane is large, and the CCD camera pixel size corresponds to 0.1 arcseconds of angular resolution. Although it is sometimes possible to obtain better resolution from the ground, the satellite provides continuous, uninterrupted, and uniform-quality data (see Figure 6.10), which cannot be done from the ground.

The FPP is a sophisticated package that provides high-resolution images of the photosphere and chromosphere at a number of different wavelengths. Most significant, it provides magnetic field measurements of unprecedented quality. In particular, extreme care is being taken to build an instrument that can provide not only the strength of the magnetic field, but its direction as well. This is known as measuring the "vector magnetic field," and such data may be valuable in making the comparison between the field that penetrates the solar surface and the observed structure and activity of the corona above.

CORONAS

We have seen that solar physics has a long and fruitful history in Russia and the Soviet Union. A vigorous solar physics program in Russia and partner nations continues and has led to a series of satellites devoted to the study of solar activity. Most recent is the CORONAS series, large observatories containing numerous and varied instruments to study the Sun and solar effects, including particle production and energetic emission such as X- and gamma-rays. Most recent in this series was CORONAS-PHOTON, launched on January 30, 2009, at solar minimum. Included in this observatory was a set of five instruments, collectively known as TESIS, with the project headed by Principal Investigators. Kuzin of the Lebedev Physical Institute. TESIS was designed to observe both the inner and outer corona from the solar disk out to 4 solar radii in EUV and soft x-ray wavelengths.

Solar Dynamics Observatory

The Solar Dynamics Observatory, SDO, is the first Space Weather Research Network mission in NASA's Living with a Star (LWS) Program. SDO is designed to help us understand the Sun's influence on Earth and near-Earth space by studying the solar atmosphere on small scales of space and time and in many wavelengths simultaneously. The LWS program has broad scientific objectives: to quantify the physics, dynamics, and behavior of the Sun-Earth system over the 11-year solar cycle; to improve our understanding of the effects of solar variability and disturbances on terrestrial climate change; To provide data and scientific understanding required for advanced warning of energetic particle events that affect the safety of humans; and to provide detailed characterization of radiation environments useful in the design of more reliable electronic components for air and space transportation systems. There are three major instruments included in the SDO Observatory:

- HMI (Helioseismic and Magnetic Imager):
  The Helioseismic and Magnetic Imager extends the capabilities of the SOHO/MDI instrument with continuous full-disk coverage at considerably higher spatial and temporal resolution. It also obtains line-of-sight magnetograms with an optional channel for full Stokes polarization measurements and hence vector magnetogram determination.
- AIA (Atmospheric Imaging Assembly):
  The Atmospheric Imaging Assembly images the solar atmosphere in multiple wavelengths to link changes and activity of the Sun's outer atmosphere to surface and interior changes. The assembly consists of 4 telescopes, each imaging the full solar disk at high spatial resolution in two

different UV and EUV wavelengths, selected alternatiely by a mechanism.

- EVE (Extreme Ultraviolet Variability Experiment): The Extreme Ultraviolet Variability Experiment measures the solar Extreme-Ultraviolet (EUV) irradiance with unprecedented spectral resolution, temporal cadence, and precision. This is providing fundamental knowledge of the solar EUV input to the Earth's energy balance.

SDO was launched February 11, 2010, on an Atlas IV rocket from Cape Canaveral. The latest real-time images of the Sun can be seen at http://sdo.gsfc.nasa.gov/.

Solar Probe Plus

Solar Probe Plus, SPP, will be a mission to visit the last unexplored place in the solar system, the Solar corona in the near-Sun region where the plasma is being heated and accelerated. By making direct *in situ* measurements at a distance of less than 9 solar radii from the surface, the mission will help us to understand how the solar atmosphere is heated and how it escapes the strong gravitational pull of the Sun. More than 50 years in the making, this extremely challenging mission has as its scientific goals:

- Determine the structure and dynamics of the magnetic fields at the sources of solar wind.
- Trace the flow of energy that heats the corona and accelerates the solar wind.
- Determine what mechanisms accelerate and transport energetic particles.
- Explore dusty plasma near the Sun and its influence on solar wind and energetic particles.

A drawing illustrating the Solar Probe Plus spacecraft approaching the Sun is shown in Figure 6.11. It is currently

FIGURE 6.11. The Solar Probe Plus spacecraft with solar panels folded into the shadows of its protective shield, gathers data on its approach to the Sun.

planned for launch in 2018 on a 10-year mission featuring repeated close encounters with the Sun.

Hi-C: The Quest for Resolution

We have seen that the complex interaction between the emerging solar magnetic fields and the hot coronal plasma leads to a highly structured and dynamic solar corona. In trying to answer the two basic questions about the corona – why is it so hot? and why is it so dynamic? – we are confronted with a difficult observational problem. The coronal regions that are involved in the dynamics, such as active regions and prominences, tend to be large and the magnetic configurations surrounding them are even larger. So one must build a telescope capable of imaging a volume large enough to encompass those regions. At the same time, the processes that trigger the instabilities that lead to flares and coronal mass ejections are initiated on very small spatial scales. In order to study that triggering process we need a telescope that can resolve very small structures, so that the predicted sequence of events – magnetic reconnection, formation of a thin current sheet where energy is dissipated – can be studied, to determine whether the theory is correct.

Understanding the coronal heating process also involves similar difficulties. Theory predicts that the magnetic fields will be twisted and tangled, and that energy release will involve a simplification of the field to a less complicated state, releasing energy in the process. Until very recently there has been scant evidence in the imaging data for such configurations, with the structures that we see appearing to be mostly straight and parallel to each other. One of the few remaining possibilities for retaining our theoretical explanations is that the tangled magnetic field configuration is to be found at smaller distances than had been resolved to date.

A major step toward a resolution of these questions was achieved by the flight of the High Resolution Coronal Imager (Hi-C) sounding rocket telescope, launched from White Sands Missile Range on July 11, 2012. This telescope used the EUV multilayer technology employed by the SOHO EIT, TRACE and SDO/AIA telescopes, with the difference that it is designed to achieve far higher spatial resolution than any of those instruments. As illustrated in Figure 6.12, this goal was achieved, and for the first time the braiding of coronal structures around each other was detected. More than that, even during the few

FIGURE 6.12a. The Hi-C sounding rocket telescope observed twisted and braided coronal structures during its brief flight on July 11, 2012. The braiding was observed to simplify during the few minutes of observation, for the first time providing direct evidence of such predicted processes in the corona. This image was filtered through a program that emphasizes the sharpness of the features.

minutes of the rocket flight, an unwrapping of these structures was observed, associated with flows of hot plasma along the bright threads at hundreds of km/sec. These observations offer a tantalizing glimpse of the new understanding of coronal dynamics that may be possible if such observations can be made over a longer time period from a satellite.

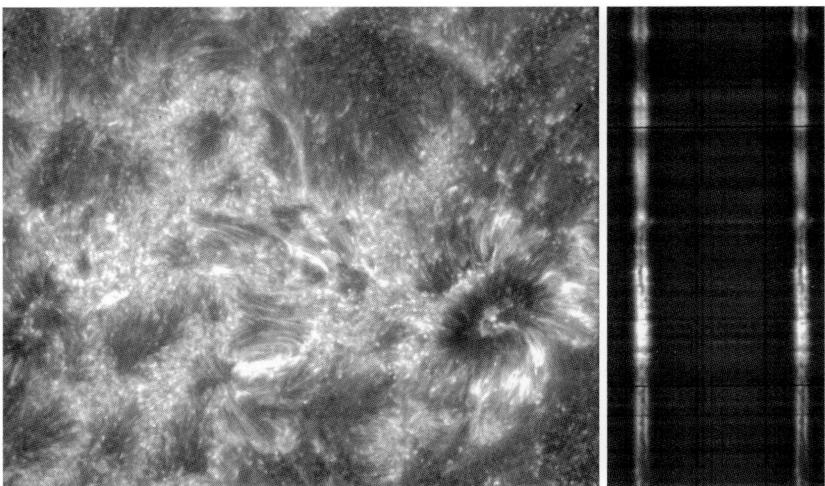

FIGURE 6.12b. First light data from the IRIS satellite, in 2013. Left: A slit-jaw image made in the 1400 Å passband. This image shows the lower transition region of the Sun. There are four passbands available for IRIS slit-jaw images. The IRIS slit runs vertically in the center of the image and has been digitally removed. Right: A spectrum of UV light passing through the slit, here showing the Mg II h and k lines at 2800 Å. IRIS can observe twelve strong spectral lines that primarily originate in the chromosphere and transition region. Further instrument specifications can be found at http://iris.lmsal.com.

IRIS

NASA's Interface Region Imaging Spectrograph (IRIS), shown in Figure 6.6, was launched on June 27, 2013. Its instrument is meant to study, in the ultraviolet, how material moves from the chromosphere through the transition region to the corona (Figure 6.12b). One of us (L.G.) built the telescope at the Smithsonian Astrophysical Observatory; the spectrograph came from Montana State University. See http://www.nasa.gov/mission_pages/iris/

SUGGESTIONS FOR FURTHER READING

Hirsh, Richard F., *Glimpsing an Invisible Universe* (Cambridge University Press, Cambridge, UK, 1983).

Lang, Kenneth R., *The Sun from Space* (Springer, Berlin, 2009).

Von Braun, W., Ordway, F. I., III, and Dooling, David, *Space Travel: A History* (Harper & Row, New York, NY, 1985).

7

Between Fire and Ice

From the point of view of determining our climate, the Earth is basically a large rock floating in cold, empty space with an enormously bright searchlight shining on one side of it. If the Earth were not rotating and did not have an atmosphere, then the side being illuminated would be hotter than boiling water and the dark side would be solidly frozen. A rotating Earth would be more evenly heated, like a rotisserie chicken being cooked, but without an atmosphere the *average* temperature around the globe would still be below the freezing point of water. The warming due to the atmosphere, known as the greenhouse effect, is necessary to keep the Earth habitable.

What we call "climate" is a complex interaction between the heating of the Sun and the processes that distribute this heat over the planet. This chapter provides a perspective on the forces determining the overall climate of the Earth. The goal is to provide the "big picture," the global climate and its relation to the Solar input as an organic whole. We will then be in a position to discuss the major causes of climate variability, and the role played by the Sun in climate change. As we will see, in the 20th century the natural changes due to the Sun's variable radiation were overwhelmed by the human influences on global climate.

THE GENESIS OF CLIMATE

By the *climate* of a region, we usually mean the average, over some long timescale, of temperature and precipitation. Since we normally omit day-to-day variability, or *weather*, in conceptualizing climate, the climate of a region might be thought of as the "averaged weather."[1] We use words such as warm or cold, moist or dry, but the inadequacy of these descriptions becomes apparent as soon as we ask a few questions: do we mean summer or winter? typical daytime or typical nighttime? do we average over a month, a year, or a century? do we mean present-day conditions or those prevailing during the last Ice Age? Is the occasional killer tornado part of the climate? What exactly do we mean by climate anyway?

The local climate of a region is determined by many factors: Is there a body of water nearby? Is the region downwind of a mountain chain? What is the pattern of ocean currents on the usual storm tracks? What is the typical composition of the soil? Is there a large urban area nearby, producing excess heat and air pollutants?

[1] Hubert Lamb, founder of the Climatic Research Institute at the University of East Anglia, and one of the forerunners of this field, considered the term "averaged weather" to be incorrect since we must specify "not only averages but the extremes and the frequencies of every occurrence that may be of interest." He suggested that climate be defined as "the total experience of the weather at any place over some specified period of time." (Lamb 1996). His objection is a real one – the range of variability in the short term at most locations is so large that averaging can be quite misleading. Moreover, the changes in "average" climate from one millennium to another, or even from one century to the next, are so large that the word "average" hardly seems appropriate. Yet one has the intuitive feeling that the word "climate" has meaning, and we will for the time being speak as if "typical climate" is a meaningful concept.

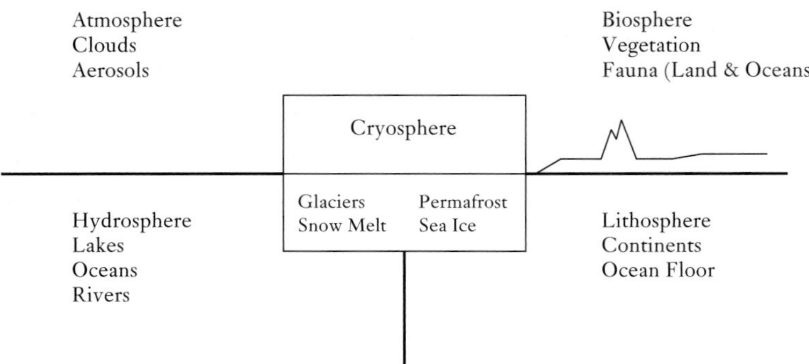

FIGURE 7.1. Schematic illustration of the climate system and its subsystems.

What we really need to speak about is the *climate system.* Climate is determined not only by what the air is doing, but also by the interaction of air, sea, land, and ice, forming one interconnected whole. Changes in one part affect the others and are in turn affected by them. What we have is a complex, dynamic linking of the atmosphere, the oceans, lakes, and rivers (hydrosphere), the frozen water masses (cryosphere), the underlying ground (lithosphere), the plant and animal life (biosphere), and the energy source that drives the interactions among all of these, the solar radiation. These systems are all interlinked by exchanges of energy, mass, and other dynamic variables to form a local climatic system, as illustrated schematically in Figure 7.1.

The basic mechanism behind all of these dynamics is differences in heating between one place on the Earth and another, followed by a redistribution of energy from warm regions to cool ones. The ultimate source of nearly all the energy driving the climate system is the Sun, and one of the goals of climatologists is to come up with a model that takes this energy input, feeds it into a model Earth, and comes out at the end with the observed climate. Because the real system is quite complicated, this goal has not yet been achieved fully, but we can approach it in stages, starting with the most fundamental parts and then adding to the model bit by bit. In mathematical language, this is equivalent to starting with the largest terms in the equations and then adding additional terms as we become more familiar with the solutions.

A simple climate system model

First, imagine some surface exposed to the Sun – the hood of your car on a sunny day might be a good example (see Fig. 7.2). Energy from the Sun, mostly in the form of visible and infrared light, is constantly falling on this surface. Some of this energy is absorbed into the surface, causing it to get hotter. We now ask, how hot will it get?

If the surface just keeps absorbing energy it will keep getting hotter forever. Since we know that this doesn't happen, there

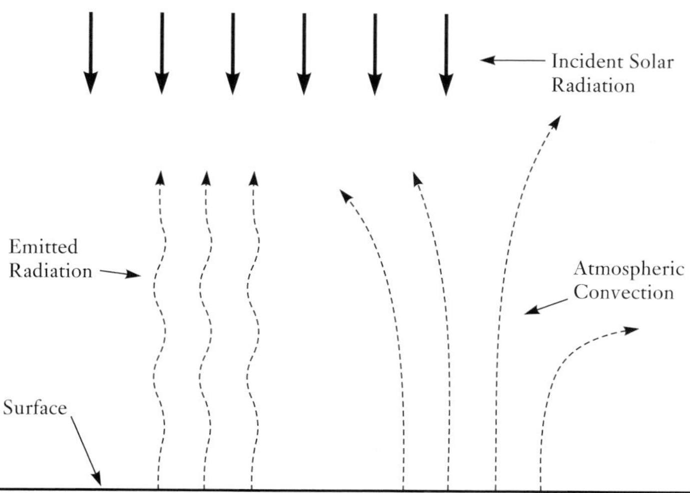

FIGURE 7.2. Heating of a surface by the balance of incoming and outgoing energy.

must be some process or processes that remove energy from the surface. In this case, the two major ones are: 1) air in contact with the surface is heated and then moves away, taking some of the energy with it, and 2) the surface radiates energy into the surrounding space. (There are also other processes operating, such as conduction of heat energy into and along the metal, that we can omit for the simplicity.) The first process is familiar and needs little explanation: anything in contact with the heated surface will tend to come to the same temperature as that surface. In particular, the air around it will be heated, and movements of the air will transport some of that heat away. Note that this implies that in a vacuum the same surface, exposed to the same illumination from the Sun, will be hotter than in air. This is indeed true for objects in space or on the Moon.

The loss of energy by radiation is somewhat less familiar to most people. All objects radiate electromagnetic energy (e.g., light) and more or less of it is radiated, depending on the temperature of the object. At normal "room temperature" most of this radiation is in the infrared, and is therefore not visible to our eyes, but it can be detected with special infrared-sensitive cameras. Such cameras can also be used to image homes to detect heat leaks, that can then be sealed to reduce heating costs; this is especially important in cold climates. The reason that all objects emit such radiation is that matter is made of atoms, which consist of charged particles – a positive nucleus surrounded by negative electrons – and that charged particles emit radiation when they are accelerated.

What does acceleration of charged particles have to do with temperature? What we really mean by temperature is the motion of the atoms (or molecules) in a quantity of matter. High temperature simply means a rapid motion of the atoms. The understanding of this relation between such gross, averaged quantities as "temperature" and the detailed states of the numerous, submicroscopic particles of which the matter is composed, was one of the greatest triumphs of late-19th century physics. (These studies also led to quantum mechanics in the 20th century.) Temperature corresponds to rapidly moving atoms, which also means atoms colliding with each other, since numerous atoms moving around at high speed will unavoidably encounter one another from time to time. In some cases, when atoms are held in place so they cannot move large distances, the energy takes the form of vibrations or other periodic motions such as rotation around the fixed spot.

All of these situations represent types of accelerated motion, since only motion at a constant speed in a straight line is unaccelerated. "Acceleration" means "change of motion," so that deflection via collision, or oscillation in place, represent examples of accelerated motion. Typically it is the very lightweight electrons that are most easily moved and therefore radiate most efficiently.

The amount of radiation emitted by matter increases very quickly with increasing temperature. Qualitatively, the rapid increase is understandable first because, as the temperature goes up, the atoms are moving more quickly so they collide more often. The collisions are also more violent, i.e., the accelerations are greater and larger acceleration means that the radiation is stronger. It also means that the typical photon of light emitted in the collision is more energetic, i.e., shorter wavelength.

Higher temperature thus leads to much stronger, shorter wavelength radiation. While objects at what we call "room temperature" emit their photons mainly in the infrared, for the much hotter Sun the radiation peaks in the visible (or more likely, it is "the visible" because this is where the Solar radiation peaks). Hotter stars are bluer in color because their peak photon energies are even higher.

Thermal balance

We can now understand why the surface in our example reaches a certain temperature and then does not continue to get hotter. As it absorbs energy from the Sun, its temperature increases. But conduction to the surrounding air and radiation into the surrounding space both increase with the temperature of the surface. Eventually, the surface becomes hot enough that it is losing as much energy to these two processes as it is gaining from the incoming sunlight. It has achieved *thermal balance* and will remain at a constant temperature so long as nothing else changes, e.g., a passing cloud dims the Sun, or the air becomes humid, thereby changing its energy-carrying capacity.

A simple calculation of the Earth's equilibrium temperature based on the above picture (including the albedo of the Earth, i.e., the fact that some of the incoming Solar radiation is reflected back out) gives about 255 K (0°F), whereas the correct value is about 33 K (60°F) higher.[2] Clearly there is more involved than has been included in our model so far, otherwise the Earth would be completely frozen over.

Heating a spherical rotating Earth

The simple picture of a car hood above would lead to a situation in which the surface is heated to some temperature and then remains there. While this represents the basic warming mechanism for the Earth as a whole, there are at least three additional factors that have to be considered in trying to model the Earth's climate:

- The Earth is a spherical body.
- The Earth is rotating.
- The Earth has an atmosphere.

[2] K stands for kelvin, a unit of temperature measured from absolute zero upward. On this scale water freezes at 273 K, which corresponds to 0°C or −32°F; water boils at 373 K, or 100°C, 212°F.

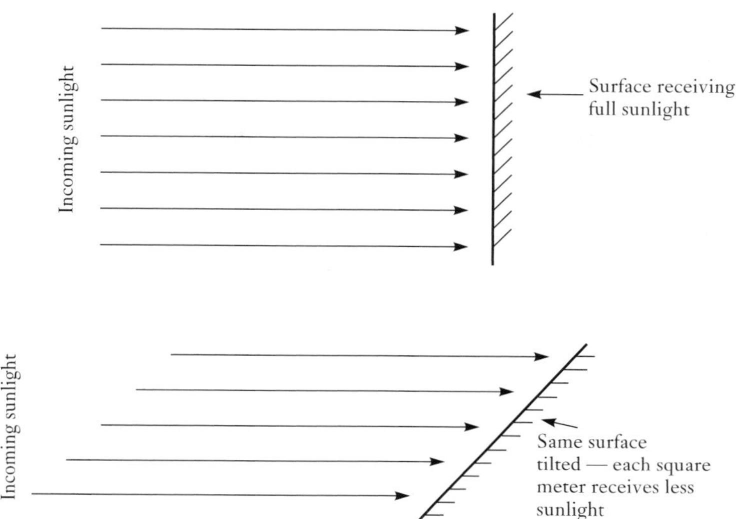

FIGURE 7.3. For a spherical Earth, different latitudes have different tilts relative to the incoming sunlight, causing differences in effectiveness of Solar heating.

Sphericity. Because the Earth is spherical, the latitudes near the equator face the Sun more directly and so receive more heating than do those near the poles, as shown in Figure 7.3. This means that the air in the tropics will warm and expand, while air in the polar regions will cool and contract. Since air is free to move, the heated air near the equator expands, and moves outward and upward. Air from the equatorial regions then flows away from the equator poleward in both Northern and Southern directions. This flow eventually cools and sinks, then flows back down toward the equator as a surface flow.

The overall pattern is two torus-shaped rolls, one in each hemisphere. This type of circulation, as a pair of giant convection cells between pole and equator, is usually called the Hadley circulation, after G. Hadley, who proposed this flow pattern in 1735. But it is only a first approximation to the truth.

Rotation. We now need to understand why these atmospheric flows are observed to go *around* the poles, rather than straight toward them. This turns out to be due to the fact that the Earth is rotating. The first effect of the Earth's rotation is, of course, the large change in solar illumination between day and night. This causes not only an overall day/night temperature difference, but also in many locations a heating/cooling of land relative to ocean, with a corresponding shift in wind direction from day to night: the land changes temperature much more quickly than does the ocean, so that in strong sunlight the land is hotter than the water during the day, and cooler than the water at night. However, this rapid cycling of the heat input averages out and has negligible long-term effect. Much more important is the effect that the rotation has on masses of air or water that try to move.

Because the Earth is a rotating sphere, objects near the equator have a large West-to-East velocity. We can easily calculate the speed by taking the Earth's circumference and dividing by the number of hours in a day: 24,800 miles divided by 24 hours is $\approx$1,000 miles per hour. Everything in contact with the surface of the Earth near the Equator – houses, people, air, birds – moves at this speed in the direction of rotation. (Rockets carrying satellites into orbit are usually launched West-to-East and close to the Equator, so that they start out with this "boost," an important factor in choosing a site on Florida's East coast for U.S. launches.)

In contrast, an object located up at one of the poles has no sidewards velocity, while intermediate latitudes have velocities in between 0 and 1000 mph. The upshot of all this is that when air flows North–South it develops a *relative* velocity that does not match that of the surface. If equatorial air moves toward the poles, it has more West-to-East velocity than the ground over which it is passing; if polar air moves toward the equator, it has less West-to-East velocity. The air will therefore curve to

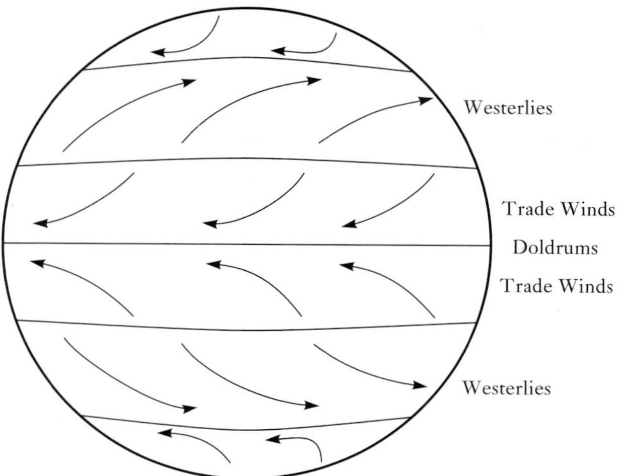

FIGURE 7.4. Typical global air circulation of the Earth.

the side, moving ahead toward the East if it is moving to higher latitudes, and falling behind toward the West if it moves to lower ones.

So the final wind flow pattern is one in which the moving air will appear to be pushed sideways, as if by a force. This apparent force, due to the rotation, is known as the Coriolis force (named after the French engineer Gustave-Gaspard Coriolis, 1792–1843, who described this effect in an 1835 paper. He also introduced the terms "work" and "kinetic energy" with their modern meanings in physics.) Figure 7.4 shows that the general circulation of the Earth's atmosphere divides into bands, with two roughly symmetric bands of high pressure air at $\pm 30^\circ$ and bands of low pressure at $\pm 60^\circ$. The winds flowing generally toward the West are called "Trade Winds" because sailors relied on them on their trade routes. Their location explains why Columbus had to first sail South in order to cross the Atlantic and therefore why he wound up in the Caribbean. The winds from the West are called "Westerlies," for obvious reasons. Near the equator the winds are not strongly curved by the rotational force, so that they are disorganized and variable. This area, known as the "Doldrums," is difficult and unreliable for sailing.

Note that there is a second band of East-to-West winds at high latitudes, above 60°. These winds were routinely used by Northern Europeans to cross the North Atlantic at least as early as the 9th century and into the 14th century. The global cooling between the time of the Medieval Grand Maximum and the Little Ice Age, described later in this chapter, caused the North Atlantic to freeze over, making crossing impossible and stranding settlers in Greenland and Newfoundland.

Atmosphere. The Earth's atmosphere acts like a blanket to retain some of the incoming Solar energy and to warm the surface. The basic mechanism of this so-called greenhouse effect is fairly simple, although as with nearly any scientific subject, the details can be formidable. The key point is that our atmosphere is not equally transparent at all wavelengths. Even though it seems invisible to us at the wavelengths to which our eyes are sensitive, the atmosphere is actually quite absorbing, i.e., opaque, at the longer infrared wavelengths. Why is this an issue? Because the global temperature is determined by the balance between incoming and outgoing energy, and the incoming energy is mainly at shorter wavelengths – those we call "visible" – so that the solar energy can enter the atmosphere and heat the surface.

But the surface, and the atmosphere near the ground, emit energy at much longer wavelengths, and the atmosphere does not allow that energy to radiate freely back out into space. This effect is sometimes called "blanketing" and it results in a warming of the Earth's surface, which leads to an increase in the strength of the radiation directed back out (and a shift toward shorter wavelengths) until a new equilibrium is achieved at a higher temperature. This type of energy trap is often generically known as a greenhouse effect. It is included in Figure 7.5, which shows an overview of the way in which the incoming solar

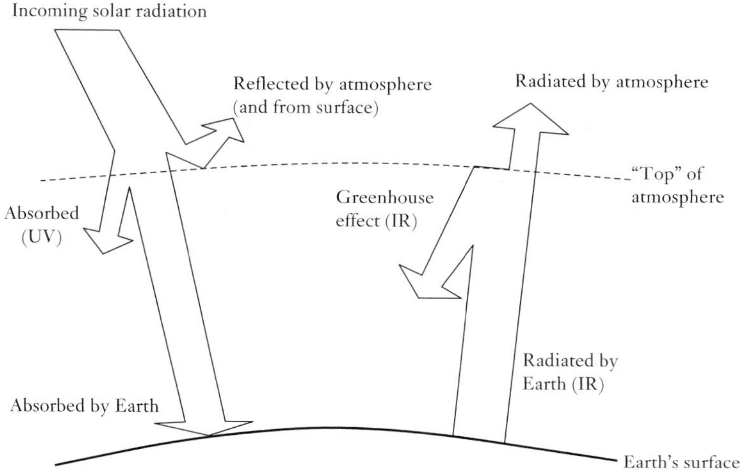

FIGURE 7.5. The energy balance of the Earth.

energy is absorbed, redistributed, reradiated, and reflected at various parts of the Earth's surface and atmosphere.

Our model is now improved, in that it shows day/night differences, it predicts that there will be a general circulation of the atmosphere, and it explains the basic greenhouse warming of the Earth. But any given place on the Earth away from the Equator also has *seasons*, so we still need an explanation for the average difference between, e.g., summer and winter. As we will see, the same factors that explain the yearly seasons also explain the far more drastic "seasons" known as Ice Ages. To make the model more realistic, we now include two properties of the Earth's orbit relative to the Sun:

- The orbit is slightly elliptical, so that the distance from the Earth to the Sun varies slowly throughout the year.
- The rotation axis of the Earth, the imaginary line connecting the North pole to the South pole, is tilted relative to the plane of the Earth's orbit. Also, the angle of tilt varies by a few degrees, with a period of about 41,000 years.

The ellipticity of the orbit means that the distance between the Earth and the Sun varies throughout the year: rather than being perfectly circular, the orbit is slightly elliptical. Because of perturbations from the other planets, the ellipticity of the Earth's orbit is not truly constant, but varies from near zero to as high as 6%, with a period of roughly 100,000 years. These days, the ellipticity is fairly low and it will remain small for the next 50,000 years; this small ellipticity may have an effect on the timing and severity of the next few Ice Ages, as we discuss later in this chapter.

The tilt of the Earth's rotation axis causes large differences between summer and winter, especially at high latitudes, and the larger the tilt, the larger the effect. Anyone who has played with a gyroscope will remember that the wheel spins many times while the axis remains fairly steady, and this is what the Earth does as it orbits the Sun. However, a gyroscope's spin axis will also slowly change in a type of motion called precession. Likewise, the direction of the tilt with respect to the long axis of the orbit's ellipse varies, like the wobbling of a gyroscope or a top, with a period of about 21,000 years. As we will see, all of these timescales show up in the climate record connected with the expansion and contraction of the great polar ice sheets.

THE CHANGING CLIMATE

By putting together the best available evidence from all sources – tree rings, isotope ratios, sedimentation rates in core samples, etc. – we can try to get some idea of the various timescales for climate change that the Earth has witnessed. The model we discussed in the previous section predicts that certain periodicities, or near-periodicities will be observed, such as day-night cycles, summer-winter changes, and the longer-term changes due to variations in the Earth's orbital parameters.

These long-term changes are indeed observed, but there are other variability timescales observed as well, on a timescale of about 200 to 500 million years. The record indicates that there seem to have been long periods in the Earth's history without ice ages, interleaved with eras having dramatic climate variations and numerous, relatively rapid ice/non-ice periods.

The key to these long-term climate changes seems to be continental drift, generalized into a theory known as plate tectonics. On a timescale of 100 million years there is considerable motion of the Earth's crust, with major rearrangements in the sizes and positions of the continents. Occasionally it will happen that a large land mass finds itself at one of the poles, making it possible for a large permanent glacier to form; this is presently the situation in Antarctica. This polar ice cap has an effect throughout the entire global climate system. In particular, it permits the orbital variations to cause periodic ice ages, through the mechanism we will discuss later in this chapter (see "The Milankovitch Cycle"). Keep in mind that what is behind all of these climate changes is a variation in the amount of solar heat retained by the Earth, and in the way this heat is redistributed among the various elements of the climate system.

First we discuss two basic issues related to the connection between the solar input and the Earth's climate: how we go about reconstructing past climates, and how we avoid concluding that the Earth should have been totally covered with ice in the past and should still be so today.

What is carbon dating?

From time to time one hears news reports of the age of some ancient artifact having been established by "carbon dating." What is this method of determining age? What is being measured? Why does this measurement tell us how old something is? To answer these questions we need to take a short detour into nuclear physics, explaining along the way why alchemists failed to turn lead into gold.

Why alchemy failed

Most of what we see happening around us in the world involves the shuffling around of electrons. Chemical reactions, the hardness or softness of materials, freezing and melting, emission of light and other electromagnetic radiation, life – these are all processes in which electrons move or are shared between atoms. Electrons are held by their attraction to the atomic nucleus, and the amount of energy needed to change or to break this bond determines whether or not such changes can happen in our typical, everyday surroundings. These changes typically require a few volts of electrical force and there is enough available in sunlight, or in the hydrocarbon burning that permits life to flourish on this planet, or in the batteries that power our cellular telephones.

But we do not routinely see lead changing into gold, although careful scrutiny does reveal an occasional atom of uranium changing into an atom of lead. But such changes are rare, and in general an element stays what it is, through fire and ice. So why do elements retain their identity, even though many other changes are happening?

The properties of elements as we see them are determined by the number of electrons in the atom. But this number is actually determined by the number of positively charged protons in the nucleus, since the nucleus will attract whatever number of electrons are necessary to make the atom as a whole have no net charge. The reason that elements retain stable properties through all of the flux of daily life is that the nuclei remain stable.

The reason nuclei remain stable is that it is a force much stronger than the electromagnetic force that is holding them together. When neutrons and protons get close enough together, they lock on to each other via the so-called strong nuclear force. Opposite charges repel, so that the protons resist being pushed together, and large amounts of energy are needed to push the protons close enough together to engage the strong nuclear force. (Think of a jack-in-the-box being pressed into its container, which then snaps closed to hold it in place.) But once the nuclei are hooked up, chemical processes do not have enough energy to affect them and to release the stored energy.

Nucleii tend to be stable at the energy levels normally encountered in our terrestrial environment. Alchemists tried to use hot fires to induce what we now know to be nuclear changes, but they would have needed a fire at many tens of millions of degrees to accomplish their goal. So they were able to achieve some superficial chemical changes, that is, a rearrangement of which elements were sharing electrons with other elements in a chunk of matter. But they did not come close to reaching the level needed to alter the composition of the nuclei, and the elements themselves remained unchanged.

Radioactive clocks

Some nuclei are more stable than others, and the ones that are not stable "decay," i.e., change, into nuclei of other elements. This decay can happen in a number of ways, almost all of which are accompanied by the ejection of a packet of energy (a photon) or some type of particle (electron, neutron, alpha, etc.) from the nucleus. Before the particles being ejected were identified, they were called "alpha ray," "beta ray," "gamma ray," for what turned out to be, respectively, a helium nucleus, an electron, and a high energy photon. If the decay involves ejection of a charged particle, then the overall charge of the nucleus has changed, and it has become a different element.

The average amount of time that passes before a nucleus decays can range from trillionths of a second to billions of years. When the elements were first created – e.g., during a supernova explosion – both stable and unstable combinations were formed. But the most unstable ones decay most quickly, so that as time passes, only the more stable elements remain. With the Earth being billions of years old, we would typically expect to find only those radioactive elements that have lifetimes of billions of years or more, and this is indeed the case. For instance, uranium-238 is relatively plentiful and it has a half-life of about 4.5 billion years; potassium-40 is also found naturally, with a half-life of 1.3 billion years. These are quite useful for geological dating, by comparing their amounts present in a sample to the amounts of their stable end-products, lead-206 and argon-40, respectively.

How is it then that carbon-14, which has a half-life of only 5730 years, is useful in archaeological dating? It should all have disappeared a long time ago, leaving only infinitesimal trace amounts in present-day organic matter. The answer is: cosmic rays. The Earth is constantly bombarded by high energy particles originating from outside our solar system, and these particles produce large showers of lower energy particles in the upper atmosphere, when they hit the first traces of air at high altitudes. Among the particles created in these showers are neutrons, which then can hit nuclei of nitrogen, which is the most abundant element in our atmosphere. As it happens, when a nitrogen nucleus absorbs a neutron, it very quickly ejects a proton, and the result is a nucleus of carbon-14 (most of the Earth's carbon is the lighter isotope, carbon-12. An isotope is a nucleus with the same number of protons, so that it acts the same chemically, but a different number of neutrons, so that the total atomic weight is different). The new carbon atom quickly binds with oxygen to form carbon dioxide, which circulates readily throughout our atmosphere, and gets incorporated into living matter.

While an organism is alive, it continues to incorporate carbon-14 along with the more common and stable carbon-12, so that the ratio of the two isotopes reflects the prevailing conditions in the environment. Once the organism dies, it ceases to incorporate additional carbon and the carbon-14 in the object is no longer replaced. The ratio of carbon-14 to carbon-12 starts to drop, decreasing by half every 5730 years. A measurement of the relative amounts of the two carbon isotopes can then tell us how long it has been since the incorporation of carbon ceased, as long as the remaining amount of carbon-14 is sufficent to make the measurement. This technique is not very helpful for geologists, who mark time in billions of years, but it has proven extremely useful to archaeologists and historians, since it can accurately establish dates back to about 75,000 years before the present.

Isotope ratios of molecules found in ice cores have provided climate data back several hundred thousand years, from the composition of the water in the ice, and also from gas bubbles trapped in the ice. We can reconstruct the temperature from the ratio of deuterium to ordinary hydrogen in the ice, and also from the ratio of a heavy isotope of oxygen to ordinary oxygen. Direct measurement of the gas in the bubbles gives the carbon dioxide concentration, and this concentration is found to correlate extremely well with the temperature, going back several hundred thousand years.

The Faint Young Sun Problem

Paradoxically, one of the problems we have with studying the Sun is that we have so much data that we can develop accurate models. As we saw in Chapter 2, the model is so reliable that we knew there was a "neutrino problem," in the sense that we were not detecting as many neutrinos as ought to be there. Similarly, the model predicts that the early Sun, billions of years ago, was much fainter than today's Sun.

The reason is this: the rate of nuclear burning depends on how closely pressed together the protons and other nuclei are, particularly down in the core of the Sun where the temperature and pressure are greatest. The rate of burning is *extremely* sensitive to this temperature, since it is mainly controlled by a quantum-mechanical effect (tunneling) that depends strongly on the separation between the particles. As the hydrogen (atomic weight = 1) of the Sun is slowly converted into helium (atomic weight = 4), the average density of the Sun increases a bit, and the core temperature goes up slightly. The higher core temperature produces a higher nuclear reaction rate, which means that more energy per second is being released and consequently the solar luminosity goes up.

The solar models predict that the early Sun was about 30% fainter than today's. Why is this a problem? Carl Sagan was one of the first to point out that the lower solar luminosity means that, for the first 2 billion years of the Earth's history, the average temperature of the Earth should have been well below the freezing point of water. However, this result cannot be correct because sedimentary rocks going back to an age of 1 billion years have been found. There is even some evidence that the young Earth was *warmer* than today. This discrepancy has been given the name "the faint young Sun problem."

The preferred solution to the problem is a large greenhouse effect. As we noted earlier, the Earth's atmosphere acts like a blanket to retain some of the incoming solar energy, so that the Earth is warmer than it would be if it had no atmosphere. The efficiency of this blanketing depends on the composition of the atmosphere, and there are many gases that can make the greenhouse effect stronger, such as carbon dioxide, water vapor, or methane. Sagan proposed ammonia, a very efficient greenhouse gas, as the agent, and this remains one of the possible answers. However, most climatologists think that it was high concentrations of carbon dioxide that were responsible, since early volcanic activity would have released enormous amounts of CO_2 into the atmosphere, and we know that CO_2 causes global warming.

The solution to the problem is still being debated. No matter what the answer, it is clear that the interaction between the Sun and Earth is quite complex and changeable – yet one more reason why we should not take for granted the present favorable climate on this planet.

CLIMATE CHANGE

There is good evidence that the Earth's climate varies on timescales of some tens of thousands of years, mainly due to the solar influence. Variations in the Earth's orbit lead to Ice Ages via changes in the amount of sunlight falling on different parts of the Earth's surface. There are also shorter-term (100-to-1000 year) climate changes that seem to be due to extended periods of higher or lower than normal solar activity levels.

It is worth noting that humans have never experienced the "normal" climate of this planet. We are now in one of the Ice Eras, when long-term polar cap glaciation is present, and this seems to be connected with the periodic recurrence of Ice Ages. Such times are rare on geological timescales, occurring roughly every half a billion years and lasting for about 100 million years. The Earth is, at most times, considerably warmer than at present and without glaciers at the poles.

On shorter timescales of one hundred thousand years or less, there have been periodic Ice Ages during the present Ice Era, the most recent of which ended about 10,000 years ago. We are now in one of the warmer "Interglacial" times, heading for another Ice Age in a few thousand years. As it turns out, we may have the good fortune to find ourselves in an unusually long interglacial period, and it may be several tens of thousands of years before the depth of the next Ice Age. This result will depend on the correctness of the Milankovitch model, discussed in the section after the next.

ICE AGES

Science is a highly dialectical activity involving a constant interchange between theory and experiment, especially in those areas where progress is being made. Where new data are involved, or a new theory is proposed, it is rarely evident immediately whether the new results are to be believed, or whether the new theory is an improvement over the old one. Those proposing change do their part by pushing hard for the new view, and those resisting change are doing their part by arguing that we have a perfectly fine explanation and don't need a new one. Without the former, there would never be any progress; without the latter, there would never be any stability.

This process is well illustrated by the 19th century debate over the existence of Ice Ages. At the start of the century, the Earth was thought to be about 6,000 years old and the evidence pointed to a great catastrophe consistent with the biblical accounts of a great flood. By the end of the century it was clear that the Earth is at least several million years old, that there have been repeated periods of extensive worldwide glaciation, and the beginnings of an explanation were being found. The story is filled with ups and downs, unusual characters, disheartening failures and dramatic successes, and it has only been in the past few decades that a generally agreed-upon explanation has emerged.

A key turning point was a lecture given in 1837 by Louis Agassiz, a young Swiss scientist, who adopted an explanation he had heard about for the presence of "erratics" – large boulders found many miles from their obvious places of origin – and the numerous gouges and scratches seen on rocks high in the Swiss Alps. One commonly accepted explanation was a flood similar to that described in the Bible. When it seemed difficult to explain the movements of boulders weighing many tons by mere water, a variant of this model involving ice rafts and boulder-laden icebergs was proposed. This was one of the more popular views at the time of Agassiz's lecture, although it carried the disturbing implication that the water worldwide had reached a height of at least 5,000 feet, since erratics were found high in the mountains. Agassiz argued instead that a massive sheet of ice had covered the region, moving boulders, carving the gouges and leaving behind massive deposits of silt as they retreated.

It took some thirty years before the view that widespread glaciation had occurred in the fairly recent past became accepted. Examination of glacial activity in such places as the Alps, the far northern UK, and the American northeast, provided convincing evidence via direct examples of the processes that needed to be explained. Glaciers can reach a maximum thickness of about one mile before the pressure of the thick layer causes the ice at the bottom to partially liquify and begin flowing. (This also applies to mountains, which can therefore not be more than about 10 miles high under Earth's gravity.) As more snow falls on top of the glacier, an equal amount of dirt and rock-laden ice is forced out of the bottom and edges. During warmer times, the edges retreat, leaving behind a characteristic type of deposit. Geologists have been able to identify the type of deposit left by each of the many processes involved and to identify the history of the advances and retreats of glaciers, eventually constructing a global map of their history since the last Ice Age.

When Agassiz was persuaded by John Lowell to accept a position at Harvard in 1847, he found that American geologists were enthusiastic about the new theory and were able to explain the sediments and polished rock surfaces prevalent in, e.g., western New York state, by glaciation. On both sides of the Atlantic, geologists were finding that the new theory helped to explain observations that had long been puzzling. A prototypical example of the way a scientist experiences a new paradigm is provided by the conversion of the great British geologist Charles Lyell to the glacier theory[3]: "On my showing him a beautiful cluster of moraines within two miles of his father's house, he instantly accepted it [the glacial theory], as solving a host of difficulties which have all his life embarrassed him." By the end of the 19th century the view that the Earth has experienced periodic Ice Ages was accepted. The problem then became one of explaining why these major climate changes occur (Figure 7.6).

THE MILANKOVITCH CYCLE

The way in which science typically proceeds is by guessing at a theoretical explanation – what Einstein called a "free creation of the human mind" – and testing the consequences of the theory against what is actually observed. Theories are not proven, they are, at best, not-disproven. The philosopher Karl Popper summarized this process as "conjectures and refutations" and proposed that a theory can only be called "scientific" if it makes predictions that can, in principle, be falsified. Put another way, a theory that cannot in principle be disproven (Popper's example was psychoanalysis) cannot be called scientific.

However, it is often easier to propose theories than to test them, so that the problem generally is one of ruling out the wrong theories, and hoping that a plausible one remains. Dozens of theories were proposed in the 19th and early 20th centuries to explain Ice Ages, most of which were shown to be inconsistent with the evidence, or were untestable and therefore left in a limbo state. It took a full century from the date of Agassiz's famous lecture before a convincing theory was formulated to explain the recurrent glaciations.

[3] From an 1840 letter to Agassiz written by the Rev. William Buckland, geology professor at Oxford. Quoted in Imbrie & Imbrie (1980).

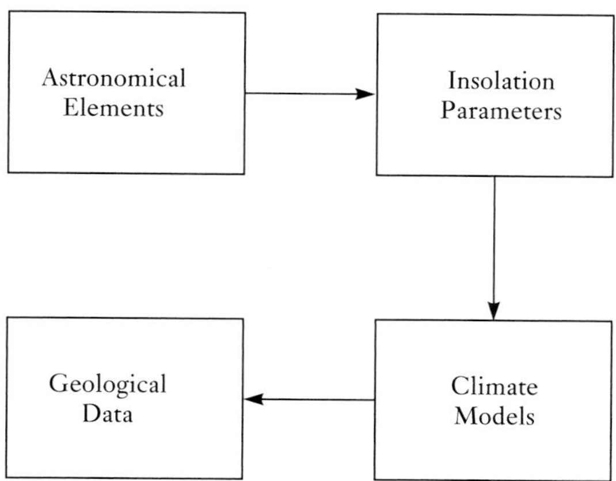

FIGURE 7.6. The four major steps in testing an astronomical theory of long-term solar influence on climate.

Ice Ages explained

The first step toward the modern explanation of Ice Ages was taken very early on. In 1842 the French mathematician Joseph Adhémar proposed that the tilt of the Earth's orbit combines with the elliptical shape of the orbit to produce an imbalance in heating that leads to glaciation. These days the Northern hemisphere is closer to the Sun in winter than in summer, but the Southern hemisphere is farther away during its winter. More subtly, the Earth moves more slowly when it is farther from the Sun and more quickly when it is closer. The result is that winters last longer in the South at the present time, and winter is when nights are long and days are short. The result is that there are more hours of darkness than of daylight in the South, and that hemisphere must therefore be growing colder due to the imbalance between heating and cooling.

This model predicts that Ice Ages will alternate between the Northern and Southern hemispheres, and that the cycle time between North and South will equal the period of precession of the Earth's tilt axis, or 21,000 years.[4] The Northern hemisphere would have been glaciated until half of a cycle in the past, about 11,000 years ago, which is indeed about when the last Ice Age ended. However, it turns out that Adhémar's reasoning was faulty, and that the total solar energy at either pole, i.e., the total number of calories it receives over an entire year, does not show the imbalance needed in this theory: any decrease during one season is exactly balanced by a compensating increase during the other season. The total amount of heat received by one pole over the course of a year is the same as the total received by the other pole.

The next major step toward an astronomical theory of Ice Ages was taken in 1864 by the self-educated Scottish geologist James Croll. The son of a stonemason, and too poor to afford university, Croll failed successively as a millwright, a carpenter, a tea shop owner, and a hotelkeeper, ending up as a janitor at a small college in Glasgow. The college had a good scientific library and Croll had ample time to read and to think about the glaciation debate. He came across Adhémar's theory and quickly realized that it was wrong, but he now added an additional factor: the amount of eccentricity of the Earth's orbit changes with time, so that the winter–summer difference is also variable, with

[4] Those familiar with astronomy might ask why we say that the precession period is 21,000 years, when it is well known to be 26,000 years. The explanation is that we are speaking about two different precession periods, one relative to the "fixed" stars and the other relative to the major axis of the Earth's orbit. The first, with period 26,000 years, is the rate of precession of the Earth's seasons relative to the stars. The latter is the one that is relevant to the issue of solar influence on Ice Ages; it is given by

$$\frac{1}{P} = \frac{1}{26,000} + \frac{1}{100,000} \tag{7.1}$$

giving P = 21,000 (in round numbers); in fact, the precession of the axis is quasi-periodic, with periods of 19,000 and 23,000 years. The period of 100,000 years is the precession rate of the Earth's orbital ellipse relative to the fixed stars, which is nearly the same as the period of variation of the orbital eccentricity. It is this latter quantity that matters for climate variability.

a period of 100,000 years. Croll reasoned that, even though the heating averaged over a year is balanced, there are epochs lasting tens of thousands of years with long, cold winters. His idea was that such seasons would cause large amounts of snow to accumulate, and he also argued that the additional snow cover would reflect back more of the incoming sunlight, so that the cooling effect would be amplified. This effect is called positive feedback and it is indeed a prominent feature of all modern climate models.

Croll went further, and realized that the patterns of air circulation, the wind patterns shown in Figure 7.4, would drive corresponding ocean currents that also transport heat from the equator to the poles. He proposed that an increase in the equator-to-pole temperature difference would increase the strengths of the winds and also cause them to shift equatorward, and that this would cause the strong westward current in the Atlantic to hit Brazil at lower latitudes, which would deflect the Gulf stream. This astounding bit of reasoning is correct, and it revealed another feedback mechanism that can also amplify the effect of a small change in solar heating on the global climate.

Croll's theory was one of the few at the end of the 19th century that fit Popper's criterion of falsifiability, since it made predictions that could be tested against the geological record. Unfortunately, despite some successes, such as the discovery that there had been periodic Ice Ages, it eventually failed this test. The theory predicts that Ice Ages will alternate between the Northern and Southern hemispheres, and that glacial epochs coincide with times of large orbital eccentricity, the last of which ended 80,000 years ago. This contradicted the geological evidence that the most recent Ice Age ended about 10,000 years ago. By the beginning of the 20th century Croll's theory was nearly forgotten.

The astronomical theory was developed in its modern form by a Serbian engineer named Milutin Milankovitch. As a young academic in Applied Mathematics in Belgrade, he decided that his talents could be applied toward solving some grand problem, and he settled on developing a mathematical theory of climate – not just for the Earth but also for Mars and Venus, and not just for the present, but for all past and future times. It took him 30 years, but he succeeded.

Milankovitch approached the problem by first calculating the planetary orbits, including eccentricity, tilt, and precession. From this he began the laborious process of calculating the "insolation" for each planet throughout the year and at different latitudes on the planet's surface. As he himself noted, had he been younger he would not have known enough to do the calculations, and had he been older he would have known better than to try.

His early publications, written in Serbian and published during the First Balkan War, went unnoticed. In 1914, during World War I, Milankovitch was captured and put in prison, but a Hungarian mathematician heard about his plight and had him released to Budapest, where he worked for the next four years. In 1920 he published his first major result, a mathematical theory for the present-day climates of the Earth, Mars, and Venus, which happened to catch the attention of a Russian-German climatologist named Wladimir Köppen and his son-in-law Alfred Wegener. The latter is remembered mainly for his theory of "continental drift" (generalized to "plate tectonics") which, like Milankovitch's own theory, would fall into disfavor and then return triumphantly many years later.

Working together, they agreed that the major change needed in Croll's theory was that it is cool summers, rather than cold winters, which lead to Ice Ages. Since winters are cold enough anyway, ice is going to accumulate, but the issue is whether or not it will melt away in the summer (averaged over many, many years, of course). If there is not more melting than accumulation, the glaciers will expand, leading to an Ice Age. The theory

Milankovitch developed is now generally accepted, although not without a 50-year struggle involving detailed geological evidence and a protracted debate over the interpretation of data from the half-million year record found in deep ice cores. The advances and retreats of glaciers, which are a measure of the changes in global climate over the past half-million years, are found to have a periodic behavior that can be linked to the periodicities in the Earth's orbital parameters: 19,000 and 21,000 years, due to the quasi-periodic precession of the Earth's tilt axis; 43,000 years, due to the oscillation in the size of the angle of tilt of the axis (which equals the "obliquity"); and 100,000 years, due to the oscillation in the magnitude of the eccentricity of the Earth's orbit. These three types of orbital changes are illustrated in Figure 7.7.

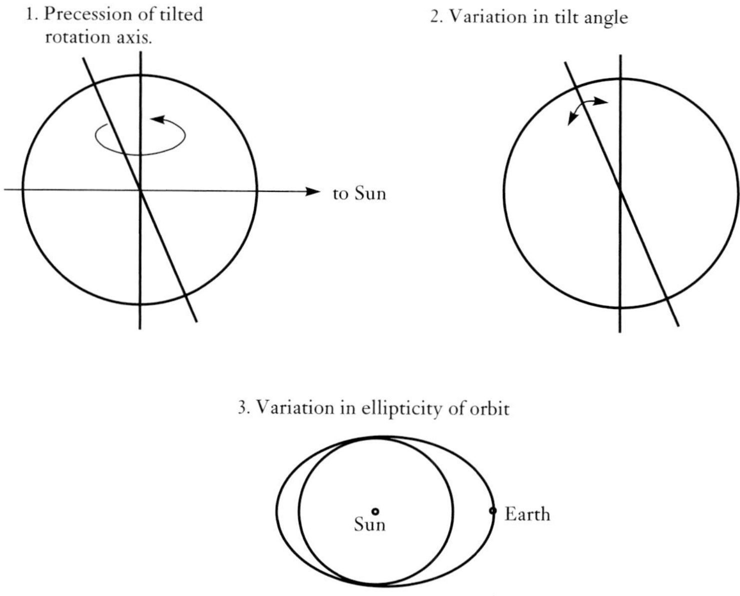

FIGURE 7.7. The three types of variation in the Earth's orbit relative to the Sun, which are thought to cause periodic Ice Ages.

GLOBAL WARMING

There is now no doubt that the Earth warmed during the entire 20th century and into the 21st. Major studies by the Intergovernmental Panel on Climate Change have carefully examined over 100 years of measurements and concluded that the average global temperature has risen about 0.8°C (roughly 1.5°F). This may seem to be an insignificant amount, until one notices that average sea levels have risen about 6 inches during that time, that alpine glaciers have retreated several miles, and that the average thickness of Arctic ice has decreased from 10 feet in 1976 to 6 feet today. Moreover, the global average temperature during the depths of the last Ice Age was only about 4–8°C (7–14°F) lower than today's, although in some places such as North America it was quite a bit colder with strong winds. Since the time of the last study, global temperatures have continued to rise and the years 1983–2012 have likely been the warmest 30 years out of the last 1400. In recent years the number of extreme weather events, from droughts to hurricanes, has increased sharply, as predicted by climate models of a warming Earth. The outbreak of tropical diseases, such as malaria and mosquito-borne encephalitis, in Northern latitudes is also being attributed to higher global temperatures and the associated changes in rainfall patterns. The rise in temperature during the past century is therefore significant, and further increases could have even greater worldwide consequences.

It is also the case that emissions of greenhouse gases, such as CO_2, have increased dramatically. While it is tempting to blame the warming on these man-made pollutants, there are some difficulties in establishing a direct connection. For instance, CO_2 levels rose steadily during the century, but global temperatures rose rapidly in the first half of the 20th century, then levelled out for two decades, then started rising again. Perhaps some storage of energy was involved (the deep ocean temperatures rose during this time), but at a minimum we know that the connection is not simple and straightforward. The basic science of warming by CO_2 is simple, and was developed by Svante Arrhenius more than a century ago, but sophisticated climate models are used in order to determine how much of the temperature rise is natural and how much is man-made, and how much is immediate and how much is delayed.

We need to factor into this discussion the effects of solar variability. There is evidence for a connection from indicators such as sunspot and radioactive carbon-14 records, but there is also evidence for a change in the relative dominance of solar effects on global warming since the start of the Industrial Age, meaning that the Sun used to be the main controller of climate change, but that man-made effects have now begun to dominate. In the following, we will examine human effects on global climate, on about 100-year timescales. Evidence for global warming, and the relative contributions of solar variations and of man-made effects are also discussed. We conclude by considering the probable future course of events.

The Grand Maximum and the Little Ice Age

There is a famous study of the relationship between the Sun and the Earth's climate that reminds us to be cautious about taking statistical correlations to be meaningful. In 1923 the British climatologist C. E. P. Brooks published the results of a study comparing the variations in sunspot number to the water level of Lake Victoria in Africa. The data covered two full solar cycles, and a comparison of the two records is quite convincing: anyone looking at the two curves would agree that there is a strong, statistically significant correlation between them.

However, no sooner had the paper appeared than the correlation between sunspot number and water level in Lake Victoria completely disappeared. Since 1923 there has been no correlation between the two. We are left wondering whether there was a real effect that became masked by some other climate change, or whether it was all merely a statistical accident in the first place.

With that caution in mind, we will describe some of the apparent connections between changes in the Sun and changes in global climate on the Earth.

One of the more productive figures in sun-climate research of the century was Charles Greeley Abbot. He lived to 101 (1872–1973) and published results over seven solar cycles – in fact, the sunspot cycle can be determined from his publication frequency. He was a powerful figure in American science, and served as the Secretary (the head) of the Smithsonian Institution in Washington until he retired in 1944. His main work, starting in 1902, only a few years after he joined the Smithsonian in 1895, was in measuring the solar constant, which he did with bolometers, instruments for measuring the total radiative power integrated over all wavelengths. By measuring the total energy emitted from the Sun year after year, he hoped to determine whether the Sun's output was variable. The main limitation to the accuracy of his measurements was in finding the variable contribution of the Earth's atmosphere to his measurements, and with hindsight we can see that he never overcame this problem, in spite of numerous efforts such as the setting up of remote stations to make observations through different bits of atmosphere in the hope of removing local effects. Abbot was convinced that he had found variations in the solar constant of 3% to 10% and that he had correlated them with the Earth's climate, a conclusion with which scientists today do not agree, because more accurate measurements from space show the effect to be less than one-tenth as large as Abbot thought.

Yet the climate record does show evidence of an 11-year influence, which would imply that the Sun is somehow driving at least some fraction of the changes. The direct variation in solar output would account for about 10% of the observed change (meaning that it would *not* account for 90% of it) unless some as yet unknown amplification mechanism exists. One recent proposal, which does not seem to work, involved cosmic ray influence on cloud formation, with the energetic particles acting as seeds for droplet nucleation in the atmosphere. The availability of more powerful computers may play an important role in answering the question, because more detailed global climate models can now be run. This is clear in a number of recent studies showing that the varying levels of ultraviolet radiation from the Sun over the course of an activity can change the atmospheric circulation patterns high in the atmosphere, which then produces a change in the atmospheric circulation even at ground level. The overall global temperature is only slightly altered, but the local climate at a given location can change significantly.

There have also been changes in climate over the past few centuries, on timescales longer than the 11-year solar cycle but much shorter than the tens of thousands of years associated with changes in the Earth's orbit. Figure 7.8 gives a rough indication of how global temperatures have varied during the past 1000 years. There are two important features to notice: during the so-called high middle ages, the climate was noticeably warmer than it is today, and starting about 1300 until the 1800s the climate was quite a bit cooler. These periods are often called, respectively, the Medieval Grand Maximum and the Little Ice Age. It seems that these little Ice Ages come and go about every 2,500 years, and there is no known cause for them. Variations in the solar activity level have been postulated, but evidence for this is scanty.

If we take the period to be 2,500 years then we are presently on the upward slope and will continue to be for another 1,000 years. This will be followed by another cooling period and will probably add to the general downward trend driven by the

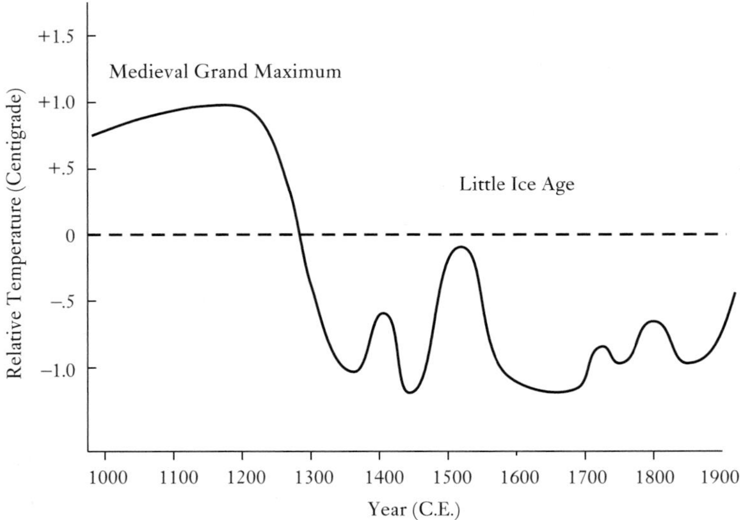

FIGURE 7.8. Schematic illustration of the global temperature variations during the high Middle Ages into modern times.

astronomical effects. The Milankovitch model predicts that the depths of the next true Ice Age will occur about 23,000 years from now.

The Industrial Age

Merely measuring the total energy coming from the Sun may not be enough to understand the ways in which the Earth's climate is affected by solar variability. There is, therefore, at least the possibility that small variations in the radiation and energy coming from the Sun can have a measureable influence on the Earth's climate. In order to test this idea, we need good measurements of the Sun's output and we need good measurements of the Earth's climate. Such data are available now, but were not available in the past. Reliable temperature data go back about 150 years, and reliable solar data go back only 20 years. To push the solar data further back, we use indirect methods, such as the correlation between sunspot number and the solar radiation output; those data go back to about 1610. For the Earth's climate in earlier centuries we can use tree rings and pollen, as well as precise measurements of certain isotope ratios in ice cores, and the organic composition of sea corals.

Any method we use becomes less reliable the further back in time we look, but we are now fairly certain that we have good determinations for the past 1000 years. The result is striking: up to the 19th century, the changes in the Sun and the changes on the Earth correlate very well, meaning that changes in the Earth's climate could have been due largely to solar changes. In the 19th century there were some exceptions due to major volcanic eruptions that cooled the Earth for several years by throwing an obscuring veil of dust high into the upper atmosphere. The correlation in those years is straightforward between the global climate and these rare and unusually large terrestrial events. In the 20th century – that is, since the Industrial Age – there have not been any such enormous volvanic eruptions, but the correlation with changes in the Sun has nevertheless broken down. It seems that the large, rapid rise in temperature and other changes in the Earth's climate can no longer be attributed to natural effects from the Sun, but are due mainly to human activity ("anthropogenic forcing"), such as increased carbon emissions into the atmosphere (Fig. 7.9).

IMPLICATIONS

There is very little doubt that the increased emission of industrial pollutants such as CO_2 is causing global warming, which will continue as long as the CO_2 levels continue to increase. But it is not at all obvious that the consequences of human activity have shown themselves fully to date in the global climate system. The linkage of chemical and biological processes involved in the worldwide "carbon cycle" is extremely complex, involving the

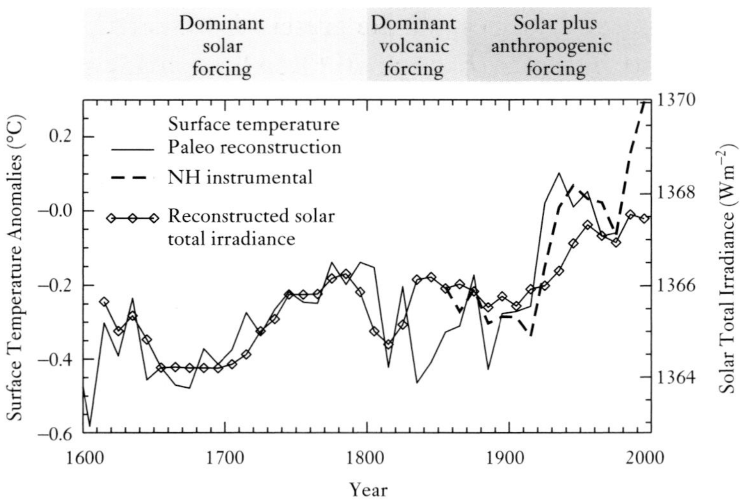

FIGURE 7.9. Comparison of solar irradiance with temperature at the Earth, from 1600 to the present. Significant volcanic events kept global temperatures low in the 19th century, but the temperature ("NH") has risen more steeply than has the solar input for most of the 20th century. It has risen especially quickly in the past two decades, while the solar activity level has remained nearly unchanged or even decreased slightly.

atmosphere, the oceans, plant and animal life, and the soil and sediments. The oceans, for instance, absorb CO_2 but become more acidic in the process. The ecosystem as a whole can add or remove carbon to or from the atmosphere, but our best estimates indicate that atmospheric CO_2 will double by the middle of this century compared to the level in 1950. In addition, the residence time for CO_2 in the atmosphere is about 1000 years. This means that even if we stopped adding CO_2 today, we would be stuck with present-day levels for many centuries, while the effect of the increased levels takes decades to show fully and is only now beginning to become manifest.

What would be the consequences of a doubling of CO_2 in the atmosphere? The average during the Little Ice Age from the 14th to the 19th centuries was only 1°C lower than today's, and the warming at the peak of the present Interglacial, about 4,000 to 8,000 BC, was only about 1°C above today's average. A change of a few degrees is therefore large and significant in terms of its effect on the global climate. Equally as important, higher global temperature means increased evaporation, but also increased ability of the atmosphere to hold moisture. The net result is an amplification of the local conditions, either drought or flooding, and to larger, more powerful storms. We had a small taste of this effect in 2011 and 2012, which set records for the number and variety of extreme weather events that occurred. Global warming therefore means an increase in the number of devastating natural disasters. As if this were not enough, water vapor is itself a greenhouse gas, so that a rise in global temperature leading to increased evaporation can itself lead to a further rise in global temperature, which causes even more evaporation, and so on.

It might be tempting to argue that, since the world is now undergoing a gradual decline in temperature based on the Milankovitch theory of ice ages, the man-made warming may prevent us from descending into another ice age. But there are several problems with this reasoning. First, the timescales involved are very different: the next ice age is coming, but it is thousands of years away, whereas the global warming due to fossil fuel burning is arriving very quickly, within a few decades. Human activity might then cause an enormous upswing in global temperature followed by a more drastic downturn than would otherwise have happened. Moreover, the warming which is now either underway or imminent has not been intentional, but rather a side-effect of our efforts to produce energy for an industrial economy. Given our present rudimentary understanding of global climate, it would be difficult to produce a controlled, planned change. The likelihood that an unplanned, uncontrolled change would be beneficial is extremely low.

A United Nations Framework Convention on Climate Change was succeeded by a binding protocol, adopted in Kyoto, Japan, in 1997, which took effect in 2005. The Kyoto Protocol set binding targets for 37 industrialized countries and the European community, though only allowing a 5% growth of greenhouse gases over 1990 levels during 2008-2012. At the Doha Climate Gateway meeting in 2012, the Kyoto Protocol was extended until 2020. A stronger treaty with more global application is being negotiated until 2015, to take effect after 2020.

Worldwide, the emission of carbon into the atmosphere is being contained a bit by the movement away from coal (except in China) to oil to natural gas, which represents a trend toward lower carbon per unit energy produced. Moving to hydrogen as a fuel can entirely eliminate carbon emission, since it does not contain any carbon.

SUGGESTIONS FOR FURTHER READING

Williams, J., *The Weather Book*, 2nd. ed. (Vintage Books, New York, NY, 1997).

Imbrie, John, and Imbrie, Katherine Palmer, *Ice Ages: Solving the Mystery*, 2nd ed. (Harvard University Press, Cambridge, MA, 1980).

Website for the United Nations Intergovernmental Panel on Climate Change (IPCC): http://www.ipcc.ch/

Emanuel, Kerry, *What We Know About Climate Change* (MIT Press, 2012).

Archer, David and Rahmstorf, Stefan, *The Climate Crisis: An Introductory Guide to Climate Change* (Cambridge U. Press, 2010).

Website for the on-line journal Consequences: http://www.gcrio.org/CONSEQUENCES/

8

Space Weather

Several of the earlier chapters in this book have mentioned how safe and sheltered we are, living on the surface of the Earth beneath a protective blanket of atmosphere and magnetic field. Conversely, when we venture outside of our safe haven, we step into hazardous territory. "Empty" space is actually far from empty: it is filled with high-energy particles and radiation, bullets of matter shooting in all directions, clouds of hot plasma thrown out by the Sun, and extremes of heat and cold (simultaneously). Not all of the hazards come from the Sun: on February 15, 2013, a meteoroid – a rock from elsewhere in the solar system – exploded in our upper atmosphere and sent shock waves across central Russia so strongly that 4000 windows were blown out, injuring over 1000 people with the fragments of glass and otherwise. Many thousands of smaller objects hit the Earth every day, but again our atmosphere protects us from all but the largest ones.

The Sun is one of the major sources of energetic particles and radiation which affect the Earth. The solar corona, even when it is not producing a major eruption, is so hot that the enormous gravitational pull of the Sun cannot completely contain it. The result is an expansion of its million-degree plasma into interplanetary space to form a "solar wind," which roars outward at supersonic speeds of hundreds of miles per second. Rapid, intense explosions in coronal active regions, known as "solar flares," produce bursts of x-rays, gamma rays, and high-energy particles at the Earth. Large-scale reconfigurations of the magnetic field of the Sun cause "coronal mass ejections" that send enormous clouds of hot ionized plasma and magnetic fields toward the Earth, often with severe consequences.

Our small planet glides innocently through these perils at about 19 miles per second in its yearly orbit around the Sun. Its thin atmosphere provides some protection from the particles, radiation, and debris, and its weak magnetic field puts a shield-like barrier out into space which deflects all but the most energetic charged particles. Inevitably, though, the portions of our upper atmosphere that absorb the radiation are affected by it, and the Earth's magnetic field is disturbed and distorted by its interaction with the solar wind and the mass ejections. This kind of dynamic activity in Earth's vicinity produces a host of effects which are now generically referred to as "space weather."

Weather forecasts

In the same way that weather forecasts are issued by the National Weather Service, the Space Weather Prediction Center (SWPC) in Boulder, Colorado issues space weather alerts, watches, and warnings.[1] The SWPC defines these three levels of announcement as:

1. Watch: A watch is issued when conditions are favorable for a certain level of activity to occur.

[1] The SWPC is part of the National Weather Service, which is, in turn, part of the National Oceanographic and Atmospheric Administration (NOAA).

2. Warning: A warning is a high-confidence prediction of imminent activity.

3. Alert: An alert provides rapid notification that an activity-detection threshhold has been reached.

The function of the SWPC, is "Providing space weather alerts and warnings to the nation and the world for disturbances that can affect people and equipment working in space and on Earth." The format of these alerts is similar to that used for hurricanes, including a 5-level organization of geomagnetic storm strengths as is done for hurricanes. Information about current conditions is available at the SWPC Website, http://swpc.noaa.gov. In this chapter we examine the solar conditions that cause these disturbances.

THE SOLAR WIND

By the middle of the 20th century, it was becoming clear to solar physicists that there is more than light and heat coming from the Sun and influencing the Earth. Both strong auroras and disturbances of the Earth's magnetic field were seen to have a recurrence period of 27 days, which is the same as the rotation period of the Sun as seen from Earth. It was also known that auroras and magnetic disturbances are stronger near the Earth's poles, which indicates that whatever is coming from the Sun is being funneled up into the polar regions. The obvious assumption – for a physicist – is that charged particles, electrons, protons, and ions, are being deflected toward the poles by the Earth's magnetic field, and that they are being spewed out from some location on the Sun. This emitting region on the Sun apparently can last for at least a few solar rotations, since some of these disturbances are recurrent. Although a name was given to these hypothetical areas – "M" regions – by Julius Bartels in 1932, the source of the magnetic storms could not be found on the Sun. Several different lines of work had to come together before the explanation was found.

During the early 1950s, the German theoretical physicist Ludwig Biermann published a series of papers on comet tails. He argued that the usual explanation for why comet tails always point away from the Sun – radiation pressure, from the sunlight falling on the comet tail particles – was grossly inadequate to explain the observations. Instead, he proposed that a stream of high-speed charged particles flowed out from the Sun. He even calculated the speed, 300–600 miles per second, and the number of particles in the flow, roughly 1000 ions and electrons per cubic inch.

Not everyone agreed and the theory was not accepted for many years. Meanwhile, a mathematician turned solar physicist named Sydney Chapman decided to calculate how far out from the solar surface the corona should extend. His inspiration for doing this calculation was the realization that a hot plasma has many free electrons in it, and they are extremely efficient at carrying heat. This means that energy from a hypothetical coronal "base" near the surface of the Sun can be transported rapidly outward into the upper parts of the corona. The result is that, instead of the temperature dropping rapidly as one proceeds outward, it falls very slowly. For instance, if the coronal temperature is 1 million degrees very close to the Sun, then it is still 200,000 degrees at the distance of the Earth. The corona is greatly extended, although sparsely populated at 1 au (au stands for Astronomical Unit. It is equal to the average distance between the Earth and the Sun, 8.3 light minutes, and was defined at the General Assembly of the International Astronomical Union in 2012 as exactly 149, 597, 870, 700 m). We might even say that the Earth is in the corona.

Chapman's calculation was done for a static corona, in which the gas extends upward to some height because of its temperature, but which basically is held in place by gravity. In 1957, as he was about to take a position at the University of Chicago, the physicist Eugene Parker began working on the problem of explaining the solar cycle-related variation in cosmic ray intensity. Cosmic rays are high energy charged particles, mainly protons, that fill interstellar space and constantly bombard the Earth. But it is found that the number of cosmic rays varies with the solar cycle, being lowest near solar maximum, particularly for the weakest (lowest energy) particles. It could be that a high rate of particle and magnetic field outflow from the Sun during years of high activity helps to push back the incoming cosmic rays, so that fewer of them reach Earth when the Sun is active. Parker discussed the possible role of the Sun with both Chapman and Biermann, and soon realized that he had to find a way to reconcile Chapman's static but extended corona with Biermann's outflow of material from the Sun.

The key turned out to be in the far reaches of the solar system, out beyond the orbits of Jupiter and Saturn, where the solar system merges into the space between stars, the so-called interstellar medium. This region has a very low density of particles, and Parker noted that Chapman's model did not have the corona diminishing quickly enough: his model predicted a fairly large gas pressure at the junction with the interstellar medium, far higher than what is actually out there. (The word "large" is a relative term here. The pressure was supposed to be 0.00000000000000000000001 pounds per square inch, but it was actually calculated to be more than 0.0000000000000000001 pounds per square inch.) Parker considered the possibility that Biermann's outflow of material was due to a continual expansion outward of Chapman's extended corona.

By solving the relevant equations, he found that a unique solution for an outflowing solar wind existed that was physically realistic (in particular, that started at the surface of the Sun with a low, rather than a high, velocity) and that also had a low pressure at infinity (which is the technical term for "very far away"). This solution, for a coronal temperature of a few million degrees near the Sun yields an outflow speed at 1 au of about 1000 miles per second and a particle density of a few thousand ions per cubic inch, just what was required by Biermann's calculations.

The theory was highly controversial and was not well-received. It was not until the 1960s, when spacecraft actually measured the particles flowing out from the Sun, that the high-speed solar wind was accepted as real.

THE EARTH'S MAGNETOSPHERE

It has been known since the publication of William Gilbert's *De Magnete* in 1600 that the Earth has a magnetic field, as if it were a giant magnet (the full title of his work is "On the Loadstone and Magnetic Bodies and on the Great Magnet of the Earth"). The work was based on many years of observation and experiment with iron-rich magnetized rocks, or lodestones, and it was motivated in large part by the need to explain to mariners why their compasses worked, and also why they sometimes failed to work. In any case, we know that the Earth is surrounded by a large shell-like invisible magnetic field, whose existence is inferred from its effect on things like magnetized needles. Even in the 17th century it was known that the location of the magnetic pole was changing, and that motion has continued to the present day (Fig. 8.1).

This field extends far out above the Earth's surface and into space (Fig. 8.2). As indicated by the direction of the arrows in the figure, the north magnetic pole of the Earth is in the geographical south these days. There are reversals of the Earth's field at random intervals averaging a few hundred thousand years or so. If there were no disturbances from the Sun, the shape of the field would be very simple and not terribly interesting, as shown

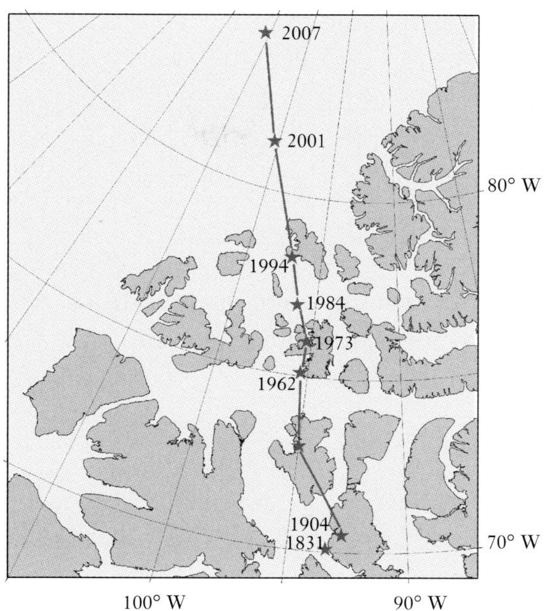

FIGURE 8.1. The location of the Earth's magnetic North pole (and the South pole as well) has been moving over the years. The strength of the field has also decreased with time, possibly indicating that a reversal of the field is imminent.

in the figure. However, the Sun sends out a wind of energetic ionized particles, as discussed above. As this wind blows past the Earth, it produces a profound distortion of the Earth's magnetosphere – the magnetic volume surrounding the Earth – as shown in simplified form in Figure 8.3. The dynamic interaction between the solar wind and the magnetosphere changes the simple, dull, bar-magnet pattern into one rich in complications and in hazards to man and machine, both in space and on the ground. The main effects are:

1. The Earth's magnetosphere deflects the solar wind, which then follows a complex route around and partly into the vicinity of Earth;

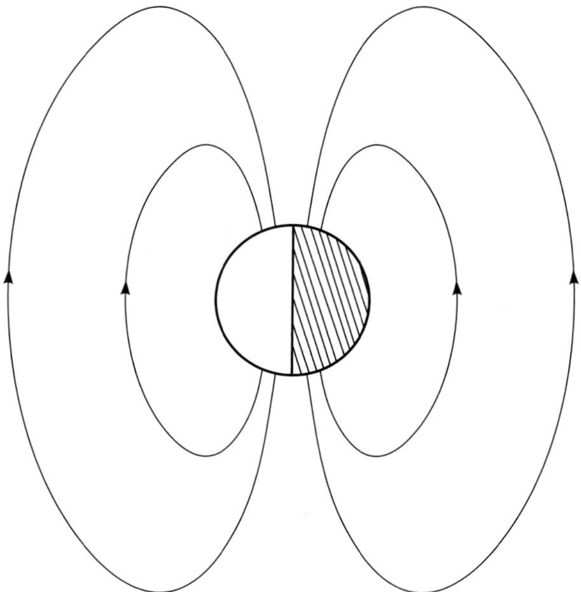

FIGURE 8.2. The Earth's magnetic field without solar wind effects.

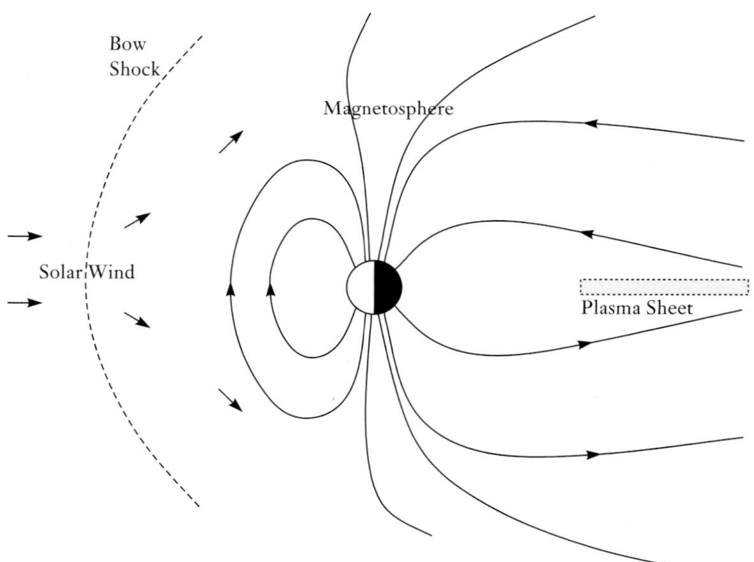

FIGURE 8.3. The Earth's magnetic field disturbed by, and interacting with, the solar wind.

2. The solar wind carries some of the Sun's magnetic field with it, which opens up and reconnects with a portion of the Earth's field;

3. A huge current circuit, carrying over 1,000,000 amps, flows from the magnetosphere through the upper atmosphere and out again, with an associated auroral display (Plate XIII);

4. A long tail of highly dynamic plasma forms in the anti-Sunward direction, and is subject to instabilities that propagate back and cause dramatic effects at Earth;

5. Energetic charged particles become trapped in torus-shaped "radiation belts" near the Earth.

Auroral particles and global currents

The auroras,[2] or Northern Lights, are spectacular displays consisting of glowing curtains of light shimmering high above the ground and lasting for many hours or days. It has been known for decades that the Sun somehow causes these displays, and for many years it was thought that energetic particles from the Sun impinge directly down onto the Earth's atmosphere, releasing their energy as they strike the first traces of air (Plate XVI). The mechanism responsible for the light show is ionization – the excitation of electrons in an atom to a higher quantum level or their expulsion out of the atom entirely – of oxygen and nitrogen atoms, followed by a recombination of an electron with the ionized atom or the return of an electron from a higher level to a lower level in the atom. These processes lead to the release of energy from the atom, in the form of a photon, and it is this light that we see as the aurora.

We now know that this simple and direct view is not correct, and that auroras are caused by a somewhat more complicated process. The flow of energetic charged particles from the Sun

[2] This should be "aurorae," but we are giving in to the pressure of common usage.

acts as a generator, via the same basic process involved in a man-made generator: moving a conductor through a magnetic field induces a flow of charged particles in the conductor. In this case, the conductor is the high-speed plasma from the Sun blowing past the Earth, and the magnetic field is that of the Earth – the magnetosphere – extending out high above the Earth's surface.

A flow of current in one way or another involves a circuit, with the current flowing along a complete, closed conducting path. In the case of auroras, the path of the circuit includes the upper atmosphere of our planet, a region called the ionosphere that has free electrons that can carry a current. As we mentioned in Chapter 6, the ionosphere was identified early in the 20th century from its effect on wireless communications, and it has been known for some time that auroras occur more than 60 miles above the Earth's surface, putting them at ionospheric heights. But it is only recently that the flow of currents, in the form of an "auroral oval" has been identified (Plate XIII), mainly through imaging from above by satellites. The path taken by the solar wind-induced current is down from the region of interaction between solar wind and magnetosphere at the local dawn, around the auroral oval to the local dusk, and back up and out into the magnetosphere. There are actually two currents, one flowing westward on the day side, and the other flowing eastward on the night side. They are known as the "auroral electrojets," and typically carry roughly 1,000,000 amps of current.

A current is a flow of charged particles, typically electrons since they are low-mass and therefore easily moved. The current flow described above, downward from the magnetosphere, around the auroral oval and back up to the magnetosphere, implies a flow of electrons down, around and back up. The downward flowing electrons deposit energy when they collide with the first few atoms of air high up in the atmosphere. These collisions excite and ionize the atoms, and the relaxation of these excited atoms back to their normal state involves the release of photons of light which we call the aurora.

The downward streams of electrons concentrate into long, thin vertical sheets, so that auroras tend to have the appearance of curtains of light. Their greenish-blue color is due to the excitation of oxygen atoms, while an occasional appearance of a light red color near the bottom of the curtain is due to excitation of nitrogen atoms. The waviness or sidewards motions of auroral curtains is not due to an actual motion of the curtain of light, but rather to a change in the place of injection of the downflowing electrons – similar to the apparent motion of water directed toward a bed of plants from a moving garden hose. A closer analogy would be the electron beam in an old-fashioned TV tube, which is swept left-right and up-down across the phosphor screen to map out the brightness pattern of the desired image.

With satellite observations now available, the auroral oval is seen to be a nearly permanent feature, varying in size and strength as the flow of hot magnetized plasma from the Sun varies. When there is an especially powerful eruption from the Sun (also having a magnetic field that is oriented in such a way that it interacts especially strongly with the Earth's field) there is a violent increase in the ionospheric currents and the associated aurora. Such an outburst is known as a "geomagnetic substorm," a rather dull name for an event that generates a million megawatts of power for several hours, ten times higher than the usual level.

Reconnection

The solar wind flowing past the Earth, carrying solar magnetic field with it, has a profound effect on the magnetosphere. In fact, it is the solar wind that confines the Earth's magnetic field to a cavity around the Earth, which we call the magnetosphere. The magnetized plasma of the solar wind compresses the magnetosphere, and a pressure balance is established between the two systems at a distance of about 10 Earth radii out, in the Sun-facing direction (the "upstream" side). On the downstream side, opposite to the Sun, the magnetosphere is stretched out into a long tail extending indefinitely far back.

The two magnetic field regions – Earth magnetosphere and solar wind plasma – are separated at a surface called the "magnetopause." The word "pause" is a generic term that can be translated "edge," just as the outer boundary of the solar wind influence in the outer regions of the solar system is called the "heliopause." In this first-level picture, the solar wind and its entrained magnetic fields slip around the magnetopause and continue on past the Earth, with very little additional interaction beyond the distortion of the geomagnetic field.

But life is not so simple. The magnetic field emanating from the Sun does not always merely slide past. Under the right circumstances, the solar and terrestrial fields interact violently, via a process called magnetic reconnection.

Figure 8.3 showed some magnetic field lines near the Earth's poles that extended out but did not close back to the corresponding region of the Earth in the other hemisphere. These are "open" rather than "closed" field lines, and they are present because of the magnetized plasma being thrown out by the Sun toward the Earth. Notice that the part of the Earth's magnetosphere that faces the Sun has a field that is oriented down-to-up, that is, from the Earth's geographic South pole up and around to the North pole. This field is normally closed – it starts at one place on the Earth and ends at another place on the Earth.

However, the solar wind brings with it some of the Sun's magnetic field, and that field may be oriented either in the same direction as the Earth's field or in the reverse direction; the latter case is shown in Figure 8.4. If the solar field is directed oppositely to that of the Earth, then an interaction between the

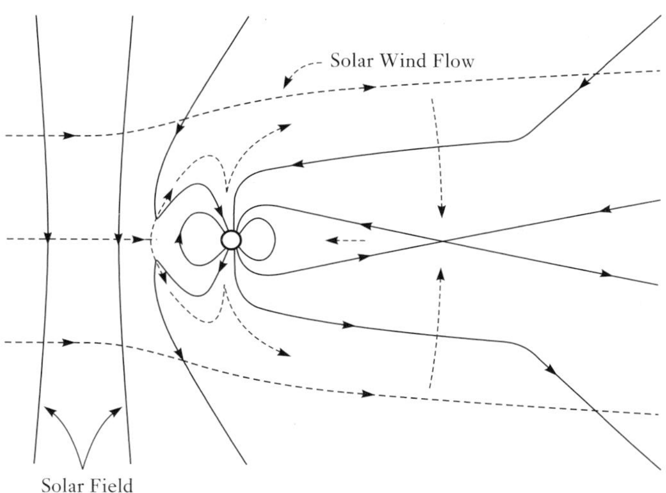

FIGURE 8.4. Interaction of the solar wind with the Earth's magnetosphere.

two can take place, in which the field lines change their connectivity. The fields can disconnect and then reconnect in a new way, as shown in the figure.

This process of breaking and reconnecting of magnetic fields is now thought to be fundamental in explaining the dynamic and highly variable solar corona and Earth's magnetosphere. The basic process involves oppositely-directed magnetic fields being pushed together, although we note that finding the "pushing" agent is a major problem in most theories. When the oppositely-directed fields come in contact they then change their connectivity, so that instead of, say, up-down field lines we end with left-right field lines. All of this action occurs in a small reconnection region, a black box within which most of the interesting and complicated physics happens. The result of this reconnection process is that jets of energetic plasma shoot out from the reconnection site; in the case of the Earth's magnetotail, this process feeds the plasma sheet which is implicated in magnetic substorms.

The reconnection process, which occurs on the dayside (the side of the Earth facing the Sun) leads to an assortment of dramatic effects in Earth's magnetosphere. First, the field lines of the magnetosphere that are involved in the opening-up process spring back toward the Earth's poles, and they bring some of the solar wind particles back with them. These field lines are then blown back toward the magnetotail, and the hot plasma (after first entering the polar regions and being bounced back out by the increasing strength of the Earth's polar fields) flows back into the tail region.

In the tail, about 100–200 Earth radii back, yet another reconnection process occurs, because the magnetic fields are again oppositely-directed. In this case, the reconnection process shoots a jet of high-energy plasma back in toward the Earth (and also outward away from the Earth), and this plasma forms the plasma sheet that was shown in Figure 8.3. Particles from this sheet drift in toward Earth and are trapped in the closed magnetic field of the interior parts of the magnetosphere, eventually forming donut-shaped rings of energetic particles known as the Van Allen radiation belts.

The Van Allen Belts

One of the first discoveries of the space age was the Van Allen belts, regions of charged particles trapped in the Earth's magnetic field. James Van Allen, with the first U.S. satellite, Explorer-1, discovered the radiation belts and found that there were at least two distinct parts, an inner and an outer belt. (The Russian satellite Sputnik-2 had also detected part of the outer belt earlier, but the significance was not immediately apparent.) Since then a number of satellites have been launched to study the radiation belts, including NASA's paired spacecraft known as the Van Allen Probes. Among other things, scientists have discovered that there are actually many separate populations in the radiation belts including various electron belts, proton belts, and belts made up of anomalous cosmic rays trapped in the magnetosphere – which Explorer-1's crude Geiger counter could not distinguish. During the first few days of operation, the Van Allen Probes discovered a transient third belt beyond the two main ones that appears to be produced by interplanetary shocks coming from the Sun.

Van Allen's two belts are now known to be the two main electron belts. The outer radiation belt extends from an altitude of $\sim 3-10$ Earth radii and consists mainly of energetic electrons from the geomagnetic tail, with additional influx from geomagnetic storms. The inner belt contains both high energy electrons and protons, and extends from $\sim 1.2-3$ Earth radii. Because the Earth's magnetic axis is tilted $\sim 11^\circ$ relative to the geometric axis, the Southern part of the inner Van Allen belt dips down close to the Earth's surface over the Atlantic Ocean, producing a region of extra particle events in low-Earth orbiting satellites when they pass through the so-called South Atlantic Anomaly, or SAA.

As the particles bounce around from pole to pole in the magnetic field they are pushed into the east-west direction by the magnetic field. Positively-charged particles – protons and ions – drift westward around the magnetic equator, and negatively charged particles – electrons – drift eastward. The net result is a huge ring current circling the Earth in the radiation belts.

FLARES

In this discussion of space weather we are interested in solar flares mainly for their effects on Earth and on the near-Earth space environment. To this end, there are two main products of these solar instabilities relevant to the discussion:

1. There are enormous eruptions of coronal material known as "Coronal Mass Ejections," which may or may not be triggered by solar flares (this is a controversial point at the moment), and which send clouds of hot, magnetized plasma out into interplanetary space.
2. In a flare, the emission of electromagnetic radiation, from radio to gamma rays, is greatly increased, and large numbers of high-energy particles are also produced.

These flare products produce a wide variety of effects at Earth: increased particle precipitation into the polar regions, dynamic variations in the Earth's plasma tail leading to geomagnetic storms, large ring currents in the ionosphere leading to ground-level power surges via magnetic induction, high levels of energetic particles impinging on spacecraft, upper atmsopheric fluctuations that interfere with communications and other electronic functions such as the Global Positioning System; and many others too numerous to list. In the next two sections we explain how these solar emissions change the Earth environment to produce this variety of harmful effects.

Figure 8.5 shows the aftereffects in the corona of a flare event, in which the coronal magnetic field erupts outward, then reconnects underneath the escaping ejection to form postflare loops. The frame shows the postflare state of an active region in the corona, seen by the SDO AIA instrument in the extreme ultraviolet (EUV) at a wavelength of 19.5 nm, corresponding to a temperature of one million kelvin. The fine structure seen in the EUV follows the orientation of the magnetic fields on the surface which have closed up following the eruption of a massive filament. The process by which magnetic energy and large amounts of hot plasma are released, followed by the closing back of the magnetic field is called "reconnection."

FIGURE 8.5. An arcade of loops formed by reconnection of magnetic fields in the corona following a magnetic eruption, May 10, 2010.

The reason that reconnection occurs is that energy is stored in magnetic fields, and when they are loaded with dense matter there is additional energy stored in the structure. This energy would normally be released fairly slowly, but when the magnetic structures become complex and tangled, it often happens that there is a simpler field geometry that represents a lower energy state. Reconnection of the magnetic field lines can then allow a rearrangement to the lower-energy configuration, and the difference in energy is released in the process. This extra energy can go into accelerating the hot plasma outward, or heating it to tens of millions of degrees, or both. It is this rapid energy release that we call a flare, and the new observations from the Hinode, SOHO, and SDO satellites are showing the way to a better understanding of these dynamic events in the corona.

Coronal mass ejections

Effects due to magnetic reconnection occur in even more dramatic fashion when the reconnection drives a coronal mass ejection, or CME, which carries more hot plasma, stronger magnetic fields, and higher energy particles than does the typical solar wind. CMEs are not easily detected from ground-based instruments, and it was only when coronagraphs were flown in space in the early 1970s that these enormous eruptions were clearly seen. A typical CME spews out 20 billion tons of coronal material at speeds of several hundred miles per second. This is roughly one-tenth of the total coronal mass, and the amount of energy released is, in standard U.S. units, up to a million trillion kilowatt-hours.

It is through CMEs that solar activity is most forcefully felt at the Earth. They are responsible for essentially all of the largest energetic proton events – in which these high-energy particles bombard the Earth and the space environment near the Earth – and they are the primary cause of geomagnetic storms (Plate XIII). One such event, observed by the LASCO coronagraph on the SOHO satellite, is shown in Figure 8.6. When such ejections are directed toward the Earth, they can produce the entire range of geomagnetic effects described earlier in this chapter. A dramatic view of the massive filament that is ejected from the Sun in these CME events is shown in Plate XIV, as viewed by the AIA telescopes on the Solar Dynamics Observatory.

The upper atmosphere

The influence of solar variability increases with height in our atmosphere, and in space the protective effects become negligible. The Earth's magnetic field continues to provide some protection at high altitudes, but it also increases the hazards by funneling energetic particles toward the poles and by storing energetic particles in radiation belts. Transient "solar

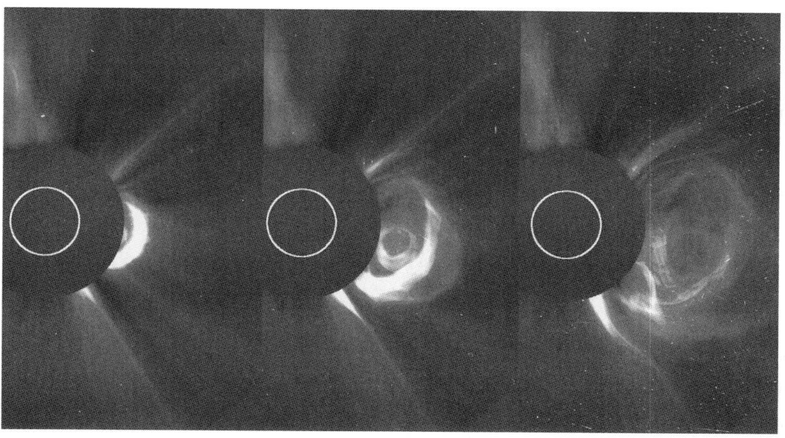

FIGURE 8.6. A coronal mass ejection, as seen by the LASCO C3 coronagraph, May 16, 2012.The CME originated low in the corona (hidden by the occulting disk covering the photosphere) and expanded outward at more than 900 miles per second. The size of the visible Sun is indicated by the white circle superposed on the occulting disk, and the spots in the right-hand image are caused by high-energy particles produced by the event.

storm" events trigger complex short-term disturbances in the magnetosphere which are felt especially strongly in the upper atmosphere and which can be fatal to both instrumentation and living beings in space. The ultraviolet (UV) radiation from the Sun has a strong effect on the upper atmosphere, producing the protective ozone layer, but also reflecting changes in the incoming solar UV intensity by changes in the observed properties of the middle and upper atmosphere.

The ionosphere

In common parlance, the word stratosphere signifies a sort of ultimate upper extreme. In actuality, the Earth's atmosphere extends a great deal further, and is classified into regions based largely on the way that the temperature varies with height:

Troposphere. The region from ground level to about 7 miles up; this is where clouds and most weather occur. The temperature generally drops with height, to about −60°F at the top of this region.

Stratosphere. From about 7 to 30 miles up, this region is relatively isolated from the troposphere, so it is usually free of water vapor and dust. Absorption of the incoming solar ultraviolet light by ozone warms the air to about 40°F.

Mesosphere. From 30 to 50 miles up, this is a region characterized by falling temperatures, down to about −90°F. Poorly explored because it is too high for aircraft or balloons and too low for satellites, so observation is difficult.

Thermosphere. Also known as the ionosphere, this is the region from 50 miles up. The incoming solar radiation has a large effect on this region, causing temperature to vary between 900°F and 3,000°F; the upper extent of the region fluctuates accordingly, and satellites in low-Earth orbit are affected by the atmospheric drag at times of high solar activity.

The large temperature changes in the ionosphere may seem surprising, especially since the solar wavelengths involved, which are mainly the UV and EUV, are only a small fraction of the total solar output. But the Earth's atmosphere is so rarefied at those altitudes that there just isn't very much of it, and it doesn't take much energy to change the temperature by a large amount. The reason it is called "ionosphere" is that the incoming solar radiation knocks electrons out of the atoms in the air, creating positively charged ions and the associated free electrons. Figure 8.7 shows how the number of electrons varies with height in the ionosphere, with one curve for daytime and another curve for night. The effect of the Sun is clearly visible between heights of about 50 and 180 miles, with the curve labelled "Noon" showing an increase of more than a factor of 100 in the number of electrons compared to the curve labelled "Midnight."

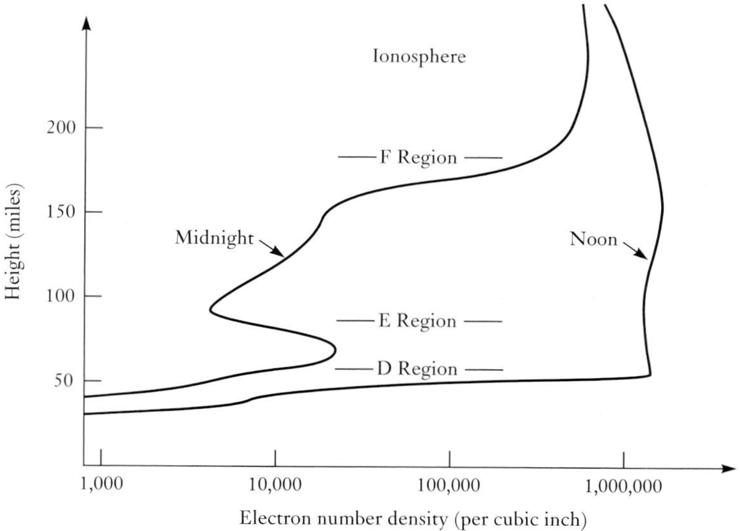

FIGURE 8.7. The number density of electrons in the ionosphere is strongly affected, on a daily basis, by the short-wavelength radiation from the Sun.

Ozone

Chlorofluorocarbons, or CFCs, are a problem that started out as a solution. They were invented in 1928 by Thomas Midgley, a chemist at General Motors, who was looking for a refrigerant to replace the toxic and explosive ammonia being used at that time. CFCs seemed ideal: the dangerous chlorine is locked up in CFC, in a non-reactive form that did not seem to break down if it was released into the atmosphere. CFC was also found to work well as a propellant (for items such as spray deodorant) and as a solvent.

It was only in the 1970s that atmospheric scientists began to investigate where all of the CFCs were going, and whether they were having any effect on our atmosphere. Soon it became clear that the CFCs continue rising high into the stratosphere, where the ultraviolet radiation from the Sun makes them break down and release the chlorine. The chlorine then feeds a complicated chain of chemical reactions that ends up destroying thousands of ozone molecules for every chlorine atom.

Why is destruction of ozone a problem? Ozone high in the atmosphere absorbs much of the incoming short-wavelength ultraviolet radiation from the Sun, thereby protecting us from these damaging rays, which can cause skin cancer and can also damage crops. Ozone is both formed and destroyed by ultraviolet from the Sun. The longer wavelength, lower energy, ultraviolet is absorbed high in the atmosphere, and it breaks apart oxygen molecules (with two atoms of oxygen) into single atoms of oxygen. These bounce around until they hit other oxygen molecules and combine with them to form the three-atom combination we call ozone. The shorter wavelength, higher energy, UV penetrates slightly lower in the atmosphere, where it breaks apart the ozone molecules, and gets absorbed in the process.

Normally, there is a balance between the production of ozone and the natural processes that remove it. But the addition of chlorine from CFCs greatly alters this balance, leading to a massive depletion of ozone. This process was first noticed in 1985 from measurements over the Antarctic, which showed that nearly half of the normal ozone was missing during southern hemisphere Spring months. Satellite observations confirmed the existence of this "ozone hole," which continued to increase in size each year. Images showing the ozone levels over Antarctica from 1979 to the present may be found at http://earthobservatory.nasa.gov/Features/WorldOfChange/ozone.php.

The mechanism for ozone depletion over the South Pole seems to be the special conditions in Winter. As temperatures fall, the high-latitude vortex of air around the Pole becomes stronger and shuts off the polar air from mixing with the rest of the Earth's atmosphere. A reservoir of chlorine forms which enhances ozone destruction, and in the Spring this ozone-poor air is released to spread over the southern hemisphere.

Beyond the particular worries about ozone over the Antarctic, these observations raise the larger question of whether this is an early warning of a more global problem. Global climate changes often show up first at high latitudes where the effects can be more extreme; this can provide us with advance indication of a spreading problem, if we are willing to heed the warning signs. In this instance, the problem was recognized, and CFC (Freon is a tradename) production has decreased in recent years. However, the residence time for CFCs in the upper atmosphere is decades, so it will be some time before the situation begins to improve.

AFTERWORD

So we end where we started, with the influence of the Sun on the Earth. The effects are seen to be multiform and complex, with potential dangers from literally all directions. We often hear it said that humans are increasingly able to control their environment, but a closer look shows that our culture has also become increasingly dependent on a controlled environment. For instance, farmers are abandoning certain types of less profitable crops that used to be grown in off-seasons because of their ability to control the growth environment. Housing is being built in locations that would have been too risky in the past – on slopes subject to mud slides, in coastal regions subject to storm damage, in desert areas where water supply is an issue. We have constructed and rely on elaborate power distribution networks, and we are becoming increasingly dependent on reliable satellite communications.

Such activities make us more dependent on our ability to retain control over the environment. To the extent that we have been assuming the constancy of the terrestrial and near-Earth climate, we are becoming more, rather than less, dependent on the constancy of the Sun. What we have shown in this book, we hope, is that this dependency requires that we continue our efforts to understand the physics of the Sun – that we develop new and better observing instruments as well as new and better theoretical models.

What is new today is that our planet's fate, as well as our ability to prosper as a human community is not only tied to the Sun, but also to our understanding of the Sun.

SUGGESTIONS FOR FURTHER READING

Suess, Steve T. and Tsurutani, Bruce T., eds., *From the Sun: Auroras, Magnetic Storms, Solar Flares, Cosmic Rays* (American Geophysical Union, Washington, DC, 2008).

Website of the Space Environment Center: http://www.sec.noaa.gov/

Pasachoff, Jay M., *Peterson Field Guide to the Stars and Planets*, 4th ed. (Houghton Mifflin Harcourt, Boston, 2000, 2012 printing).

Pasachoff, Jay M., and Filippenko, Alex, *The Cosmos: Astronomy in the New Millennium*, 4th ed. (Cambridge University Press, New York, 2014).

Appendices

APPENDIX I: GLOSSARY

ACE. Advanced Composition Explorer. A NASA-launched satellite located at the Earth-Sun Lagrange point L_1, designed to study element and isotope composition of particles, including those from the solar wind.

ACRIM. Active Cavity Radiometer Irradiance Monitor. An experiment on board several satellites, first on the Solar Maximum Mission satellite and now with ACRIM3 on ACRIMSAT, that measures the total radiative output of the Sun.

active region. Portion of the solar surface and atmosphere where strong magnetic fields, sunspots, plages, flares, and other activity are found.

angstrom. A unit of length equal to 10^{-10} meters (symbol Å).

albedo. The ratio of the amount of light reflected back from a surface to the amount of incoming light.

alpha particle. A helium atom nucleus, emitted in radioactive decay of heavy elements.

atom. The smallest unit of a chemical element; when it is subdivided, the parts no longer have the properties of a chemical element.

au. Astronomical unit, defined in 2012 by the International Astronomical Union in its General Assembly in Beijing as 149,597,870,700 m; it is roughly the average Earth-Sun distance, something first searched for from transits of Venus.

aurora. Glowing lights visible in the sky, due to excitation of atoms and molecules by downward-flowing high-energy electrons.

Baily's beads. Bright dots of light visible around the rim of the Moon at the beginning and end of a total solar eclipse or at an annular eclipse.

Balmer series. The set of spectral lines resulting from transitions to or from the first excited energy level in the hydrogen atom; compare to Lyman series.

baryon. A nuclear particle, such as proton or neutron, subject to the strong nuclear force, made of triplets of quarks.

beta particle. Also called beta ray; an energetic electron or anti-electron (positron) emitted from a nuceleus in radioactive decay.

black body. A hypothetical object that absorbs all radiation that hits it and emits radiation exactly according to Planck's Law.

blue shift. A change toward shorter wavelengths in a spectral feature; when caused by motion, it is due to a velocity of approach.

bolometer. A device for measuring the total amount of energy from an object.

bolometric luminosity. The brightness of an object taking into account all wavelengths.

calorie. A measure of energy in the form of heat; originally the amount of heat needed to raise one gram of water by 1°C.

carbon cycle. A chain of nuclear reactions, involving carbon as a catalyst in some stages, that fuse four hydrogen nucleii into one helium nucleus, with release of energy. Important in stars hotter than the Sun.

Cassegrain. A type of telescope in which light focused by a primary mirror is intercepted short of the focal point and reflected by a convex secondary mirror through a hole in the center of the primary; compare with Gregorian.

catalyst. A substance that participates in a reaction, but is left over in its original form at the end.

celestial equator. The intersection of the celestial sphere with a plane passing through the Earth's equator.

celestial sphere. An imaginary sphere of large radius, centered on the Earth, to which the stars appear to be fastened.

Cenozoic. The period beginning about 70 million years ago, characterized by the appearance of mammals.

chromatic aberration. A defect of an optical system, in which different colors are focussed at different points.

chromosphere. Part of the solar (or stellar) atmosphere between the photosphere and corona.

convection. The transfer of heat by motion of the hotter parts of a gas or liquid.

corona. A portion of the solar atmosphere characterized by temperatures above one million degrees.

coronal hole. A relatively dark region of the corona having lower density, due to magnetic field lines open to interplanetary space.

cosmic rays. High energy particles or nucleii travelling through space.

cosmology. The study of the Universe as a whole.

D_3 line. The yellow spectral line of helium, first discovered at an eclipse in 1868, near but not coinciding with the pair of D lines of sodium.

density. Mass divided by volume.

deuterium. An isotope of hydrogen, containing one proton and one neutron in the nucleus.

differential rotation. Rotation of a body in which different parts (typically different latitudes) have different angular velocities, and therefore different rotation periods.

diffraction. A spreading out of light after passing an obstacle.

diffraction grating. An object having a closely-spaced series of lines that cause different wavelengths to be spread out in space, thereby providing a spectrum of the light.

dispersion. Of light, the effect that different wavelengths are bent by different amounts in passing from one material to another. The effect is caused by differences in the speed of light for different wavelengths in a medium.

D lines. Usually refers to a pair of lines from sodium that appear in the yellow part of the spectrum.

dynamo. A device for generating strengthened magnetic fields via the motions of a conducting fluid.

dynamo theory. An explanation of magnetic field generation via a mathematical model.

eccentricity. For a planetary orbit, a measure of the flatness; for an ellipse, it is given by the distance between the two foci divided by the length of the major axis.

eclipse. The passage of all or part of one astronomical body into the shadow of another.

ecliptic. The plane of the Earth's orbit; also, the apparent annual path traced out by the Sun against the celestial sphere as seen from Earth.

electron. A particle with one electric charge, 1/1830 the mass of a proton, that is not affected by the strong nuclear force.

electron volt. The energy gained by an electron when accelerated by one volt.

element. An atom characterized by a specified number of protons in its nucleus.

emission line. Sharp energy peaks in a spectrum, caused by downward transitions from one discrete energy level to another.

energy level. A quantum state of a system, such as an atom, having a well-defined energy.

ephemeris. A listing of astronomical positions and other data that change with time; from the same root as *ephemeral*.

epoch. A subdivision of a geological period, characterized by a particular dominant characteristic.

era. A major division of geological time, composed of a number of periods.

erg. A unit of energy, corresponding to the work done by a force of one dyne (the force needed to accelerate one gram by one cm/sec^2) acting over a distance of one cm.

excitation. The raising of electrons to higher energy states in an atom.

filament. A dark, elongated feature of the solar surface seen in H-alpha; a prominence when projected against the sky.

flare. An extremely rapid brightening of a small volume of the solar atmosphere, with portions reaching a temperature of more than 10 million degrees.

flash spectrum. The solar chromospheric spectrum seen in the few seconds just before or just after totality at a solar eclipse.

flux. The amount of something, such as energy, passing through a surface per unit time.

forcing (external). A forced adjustment of the internal properties of a system via a quasi-steady state change in boundary conditions.

Fraunhofer lines. The absorption lines of the solar or stellar spectrum.

gamma rays. Electromagnetic radiation shorter than about 1 Å.

geocentric. Earth-centered.

granulation. On the Sun, convection cells about 1.6 arcsecond across on average.

grating. A surface ruled with closely spaced lines that, through diffraction, breaks up light into its spectrum.

grazing incidence. Striking at a small angle.

greenhouse effect. The warming of a planet's surface due to trapping of infrared radiation by the atmosphere.

Gregorian. A type of telescope in which light focussed by a primary mirror passes through its focal point and is then reflected by a concave secondary mirror, forming an image through a hole in the center of the primary; compare with Cassegrain.

ground level. Also ground state; the lowest energy level of a quantum mechanical system.

H line. The spectral line of ionized calcium in the ultraviolet at 3968 Å.

H-alpha. The first line of the Balmer series of hydrogen, in the red at 6563 Å.

heliocentric. Sun-centered.

helium flash. The rapid onset of fusion of helium into carbon through the triple-alpha process that takes place in most red giant stars.

hertz (Hz). A measure of frequency with units of 1/sec (per second).

Hinode. A spacecraft, formerly Solar-B, of the Japan Aerospace Exploration Agency with American, British, and European participation, launched in 2006.

H-R diagram. Hertzsprung-Russell diagram; a graph of temperature (or equivalent) vs. luminosity (or equivalent) for a group of stars.

IMAGE. NASA's Imager for Magnetopause-to-Aurora Global Exploration spacecraft, active 2000–2005.

inclination. Of a planetary orbit, the angle of the plane of the orbit with respect to the ecliptic.

interference. The property of waves that waves in phase can add (constructive interference) and those out of phase can cancel (destructive interference); for light, this gives alternating light and dark bands.

interferometer. A device that uses interference to measure properties of a source such as position or structure.

ion. An atom that is missing one or more electrons.

ionosphere. The highest region of the Earth's atmosphere, from 50 miles up; also called the thermosphere.

ISAS. Institute of Space and Aeronautical Science in Japan, now superseded by JAXA.

isotope. Of an element, one having the same number of protons, but a different number of neutrons.

JAXA. Japan Aerospace Exploration Agency.

joule. A unit of energy, equal to one kg m^2/s^2.

Julian day. The number of days since noon on January 1, 713 B.C. Noon of January 1, 2000, was Julian day 2,451,545.

K line. The spectral line of ionized calcium in the ultraviolet at 3933 Å.

Lagrange points. In the restricted three-body problem, there are five equipotential points; two are marginally stable and three are marginally unstable.

L_1. The Lagrange point between the Earth and the Sun; located about 1 million miles from Earth.

limb. The edge of an extended object, such as a star or planet.

limb darkening. The decreasing brightness of the disk of the Sun as one looks from the center toward the limb.

luminosity. The total amount of energy given off by an object per unit time.

lunar eclipse. The passage of the Moon into the Earth's shadow.

Lyman-alpha. The spectral line deep in the ultraviolet at 1216 Å that corresponds to a transition between the two lowest energy levels of the hydrogen atom.

Lyman series. A series of spectral lines in the ultraviolet corresponding to transitions to or from the ground state of the hydrogen atom.

magnetic field. A property of space that can exert magnetic forces on objects.

magnetic field lines. Directions mapping out the magnetic forces in space; the degree of packing of field lines indicates the relative strength of the field.

magnetosphere. A region of magnetic field around a planet.

main sequence. A band in the H-R diagram in which stars fall during the hydrogen-burning phase of their lifetimes.

mass. A quantity indicating the amount of force needed to accelerate an object (inertial mass) or the strength of the gravitational force produced by an object (gravitational mass); by Einstein's principle of equivalence in General Relativity, the two are equal.

Maunder minimum. The period 1645–1715, when there were very few sunspots visible.

mesosphere. A middle layer of the Earth's atmosphere above the stratosphere and below the ionosphere.

Milankovitch periodicity. The periodic variation in the Earth's climate that results from a set of variations in orbital parameters (obliquity, precession, and eccentricity).

nebula. Interstellar regions of gas or dust.

neutrino. A neutral elementary particle interacting weakly with matter.

neutron. A massive, neutral elementary particle, interacting via the strong nuclear force.

nuclear force. The strong force, one of the four known fundamental forces.

nucleosynthesis. The formation of the elements.

nucleus. Of an atom, the core, which has positive charge, most of the mass, and occupies a small part of the volume.

objective. The principle lens or mirror of an optical system.

obliquity. A measure of the tilt of the Earth; given by the angle between the Earth's equatorial plane and the plane of the ecliptic.

occultation. The hiding of one astronomical body by another.

optical. The visible part of the spectrum, 3900–6600 Å.

ozone. A molecule containing three oxygen atoms, O_3.

ozone layer. A region of the Earth's upper stratosphere and lower mesosphere where ozone absorbs ultraviolet radiation.

penumbra. a) For an eclipse, the part of the shadow from which the Sun is only partially occulted; b) of a sunspot, the outer region, not as dark as the umbra.

photosphere. The region of a star from which most of its light is radiated.

plage. The part of an active region that appears bright when viewed in H-alpha.

plasma. A hot gas in which a substantial fraction of the atoms are ionized.

plumes. Thin vertical structures in the solar corona near the poles.

pressure. Force per unit area.

prime focus. The location at which the main optical element of a telescope would form an image.

prominence. Dense solar gas at chromospheric temperatures protruding above the limb.

proton. Massive elementary particle with unit positive charge, interacting via the strong force.

period. The basic unit of geological time, during which a standard type of rock system formed; a period typically comprises several epochs and several periods are typically combined to constitute an era.

precession. The slow variation in position of the celestial sphere, due to the wobble of the Earth's rotation axis.

Quaternary. The latter part of the Cenozoic era, beginning about 2.4 million years ago and including the present.

quiet Sun. A collection of solar phenomena showing little variation with the solar cycle.

radiation. Electromagnetic energy; sometimes also energetic particles.

radiation belts. Toroidal bands of charged particles surrounding the Earth and other planets.

reflecting telescope. A type of telescope that uses mirrors to form the image.

refracting telescope. A type of telescope that uses lenses to form the image.

refraction. The bending of the direction of an electromagnetic wave at the interface between two media or in travel through a variable medium.

relativistic. Having a velocity high enough that Einstein's special theory of relativity must be employed.

resolution. The ability of an optical system to distinguish detail.

revolution. The orbiting of one body around another.

rotation. Spin on an axis.

SDO. Solar Dynamics Observatory. A NASA satellite launched in 2010.

seeing. The steadiness of the Earth's atmosphere as it affects the image quality of a ground-based telescope.

seismology. The study of waves propagating through a body, often to determine the internal properties of the body.

shock wave. A front marked by a sharp change in pressure, caused by an object moving through a medium faster than the local sound speed.

sidereal. With respect to the stars.

SMM. Solar Maximum Mission. A satellite designed to study solar flares, launched in 1980 and operational until the end of 1989.

SOHO. A joint ESA/NASA solar satellite, launched December 2, 1995, and positioned at the Earth-Sun Lagrange point L_1.

solar atmosphere. The photosphere, chromosphere, transition region, and corona.

solar constant. The total amount of energy from the Sun hitting the Earth, measured at the top of the atmosphere.

solar cycle. The 11-year variation in the number of sunspots and other magnetic phenomena on the Sun; the full cycle of magnetic variation is 22 years.

solar dynamo. The periodic generation and destruction of magnetic fields in the Sun.

solar flares. An explosive release of energy on the Sun.

solar wind. An outflow of particles from the Sun representing the expansion of the solar corona.

special relativity. Einstein's 1905 theory of invariances in transforming from one unaccelerated frame to another.

spectral line. A narrow range of wavelengths within which the intensity of a spectrum is sharply different from its surroundings.

spectrograph. A device to make and record a spectrum.

spectrometer. A device to make and record a spectrum, with more of an implication of a numerical output than for a spectrograph.

spectroscope. A device to make and look at a spectrum.

spectroscopy. The analysis of spectra.

spectrum. A display of electromagnetic radiation spread out by wavelength or frequency.

spicule. A small jet of gas seen at the limb of the Sun, about 1,000 km across and 10,000 km high with a lifetime of about 15 minutes.

STEREO. Solar Terrestrial Relations Observatory, a NASA pair of spacecraft in orbits trailing and preceding the Earth in its orbit, slightly inside and outside Earth's orbit, and so drifting ahead (STEREO-A) and behind (STEREO-B) Earth's orbital position so as first to see stereo views of solar ejections and now, with its UV cameras and coronagraphs, the far side of the Sun.

strong force. The nuclear force, the strongest of the four known forces of nature.

subtend. To take up an angle in your field of view; for instance, the Sun subtends about $\frac{1}{2}$ degree as seen from Earth.

sunspot. A region of the solar surface that is dark and relatively cool, having a strong magnetic field.

sunspot cycle. The periodic variation in the number of sunspots visible on the Sun. Typical time between minima of this number is about 11 years.

supergranulation. Convection cells on the solar surface about 20,000 km across and roughly polygonal in shape.

syzygy. An alignment of three or more celestial bodies.

TRACE. A solar satellite in the NASA Small Explorer program, launched April 2, 1998 and terminated in 2010.

thermosphere. The uppermost layer of the Earth's atmosphere; same as ionosphere.

transit. The pasage of one celestial body in front of another.

transition region. The thin region of the solar atmosphere between the chromosphere and corona.

triple-alpha process. A chain of fusion reactions in which three helium nucleii (alpha particles) combine to form a carbon nucleus.

troposphere. The lowest level of the Earth's atmosphere, in which most weather takes place.

ultraviolet. The region of the spectrum from 100–4000 Å.

umbra. Of a sunspot, the dark central region; of an eclipse shadow, the darkest part, from which the solar disk cannot be seen at all.

Van Allen belts. Toroidal regions in the Earth's magnetosphere, in which high-energy particles are trapped.

visible light. Light to which the eye is sensitive, 3900–6600 Å.

wavelength. The distance over which a wave travels during one complete oscillation.

white light. All the light of the visible spectrum together.

X-rays. Electromagnetic radiation between 1 and 100 Å.

year. The period of revolution of a planet around its central star; particularly, the Earth's period of revolution around the Sun.

Yohkoh. An ISAS solar satellite, with U.S. and U.K. participation, launched August 30, 1991 and terminated in 2004.

Zeeman effect. The splitting of certain spectral lines in the presence of a magnetic field.

APPENDIX II: THE SOLAR ENERGY SOURCE

In Chapter 2 we described nuclear burning, the process that powers the Sun and stars. In this Appendix we will discuss in a bit more detail the particular nuclear reactions that happen inside the Sun. This is particularly relevant to the neutrino problem, discussed in Chapter 4.

The p-p chain and other processes

The primary process in most stars is the so-called proton-proton chain, which starts with two nucleii of hydrogen (i.e., two protons) and ends up again with two nucleii of hydrogen plus a helium nucleus. Basically, what happens is that one proton at a time is added to a nucleus, building up heavier elements and releasing energy in the process. It is interesting to note that the steps in this reaction chain involve all of the four known forces: gravity brings the material together to produce the temperature and pressure conditions needed to begin nuclear burning; the first reaction going from hydrogen to deuterium involves the weak nuclear force; the second reaction, going from deuterium to ^3He involves the electromagnetic force; the last step, going from ^3He to ^4He involves the strong nuclear force.

In the first step of the chain, two protons fuse together to make a deuterium nucleus, which consists of one proton and one neutron. This involves the conversion of one of the protons into a neutron, which means that there has been a net change of electric charge, since the proton has electric charge of +1 and the neutron has charge 0. Since charge is conserved in this world, one unit of electric charge must disappear from the new nucleus. Indeed, it is found that a particle with electric charge of +1 is released from the nucleus in this process; it is a *positron*, or antimatter electron. Also, it turns out that a *neutrino* is produced as well. Written as a chemical reaction, the process is

$$p + p \rightarrow d + e^+ + \nu_e. \quad (8.1)$$

This reaction involves the weak nuclear force, and has an extremely small probability of occurrence (i.e., the cross-section is very small); see Chapter 4 for a discussion of the weak nuclear force. The long lifetime of stars, many millions to billions of years, is due to this "gatekeeper" reaction, whose slowness prevents stars from burning all of their fuel in only a few thousand years.

Once some deuterium has been created, the next step in the p-p chain involves adding a proton to the deuterium to make a nucleus with two protons and one neutron; formally, this is a helium nucleus but lighter than the normal helium, which has two protons and two neutrons:

$$p + d \rightarrow {}^{3}He + \gamma. \tag{8.2}$$

The γ on the right indicates that a "gamma ray," or high energy photon, has been emitted.

The last step in this particular process is the addition of one more proton to produce a helium nucleus having two protons and two neutrons:

$${}^{3}He + {}^{3}He \rightarrow {}^{4}He + p + p. \tag{8.3}$$

Note that two protons are emitted in this reaction.

There are many other possible reactions, all having far lower probability. For instance, in the last step shown above it also happens a few per cent of the time that a helium-3 nucleus hits helium-4. This initiates several further reactions which eventually end up with two helium-4 nucleii:

$${}^{3}He + {}^{4}He \rightarrow {}^{7}Be + \gamma \tag{8.4}$$

$$e^{-} + {}^{7}Be \rightarrow {}^{7}Li + \nu_e \tag{8.5}$$

$${}^{7}Li + p \rightarrow 2\,{}^{4}He. \tag{8.6}$$

Note that a neutrino has been emitted in the second reaction of this chain.

One other reaction – in which a proton hits ${}^{7}Be$ to produce ${}^{8}B$, which then decays to $2\,{}^{4}He$ with the emission of a neutrino – occurs rarely in the Sun (about 0.02% of the time) but is fairly important because it produces higher energy neutrinos, which are easier to detect at the Earth.

Although only a small fraction of the Sun's total mass is involved in nuclear burning, the number of atoms in absolute terms is large: 4×10^{38} protons per second are used up. Fortunately, the Sun has about 10^{57} protons, so that even after 5 billion years of nuclear burning only a few per cent of the Sun's hydrogen has been used up, so far.

Note that, in the p-p chain, one neutrino is produced for every proton consumed. This means that the Sun produces 2×10^{38} neutrinos per second, yielding a flux at the Earth of nearly 10^{15} per square meter per second. But, because neutrinos react so weakly, we must build huge detectors in order to measure their passage. (One of the early experiments used scrap battleship hull plates to provide the "stopping power" needed.) The experimenter then has the problem of separating out any unwanted processes, particularly those involving high energy cosmic rays, from the desired neutrino reactions. In an effort to reduce this background noise, neutrino experiments are placed deep underground, in old mine shafts or even under mountains.

APPENDIX III: SOLAR SYSTEM PHYSICAL CONSTANTS

Sun, planets, and dwarf planets

Object	Mass (kg)	Radius (km)	Radius of orbit (au)	Year (days)
Sun	1.99×10^{30}	696,000	N/A	N/A
Mercury	3.30×10^{23}	2,439	0.387	88
Venus	4.87×10^{24}	6,052	0.723	224.7
Earth	5.97×10^{24}	6,378	1.	365.26
Mars	6.42×10^{23}	3,397	1.523	686.95
Ceres	9.43×10^{20}	454 × 487 × 487	2.76	1679
Jupiter	1.90×10^{27}	71,492	5.203	4,337
Saturn	5.69×10^{26}	60,268	9.537	10,760
Uranus	8.68×10^{25}	25,559	19.19	30,700
Neptune	1.03×10^{26}	24,764	30.07	60,200
Pluto	1.5×10^{22}	1,150–1,200	39.48	90,780
Haumea	4×10^{21}	650	43.13	103,468
Makemake	3×10^{21}	750 × 715 × 715	45.79	113,183
Eris	1.7×10^{22}	1163	68.01	204,870

Photo Credits

Fig. 1.1	J. M. Pasachoff
Fig. 1.2	L. Golub
Fig. 1.3	Vassar College
Fig. 1.4	NASA/TRACE
Fig. 1.5	New Jersey Institute of Technology
Fig. 1.6	National Solar Observatory
Fig. 1.7	National Solar Observatory
Fig. 1.8	National Radio Astronomy Observatory
Fig. 1.9	Nobeyama Solar Radio Observatory, NAOJ
Fig. 1.10	L. Golub
Fig. 1.11	Jay Pasachoff, Ron Dantowitz, Nicholas Weber, and Muzhou Lu (Williams College/Clay Center Observatory/Dexter-Southfield School/NSF); composited by Pavlos Gaintatzis, Aristotle U. of Thessaloniki
Fig. 1.12	High Altitude Observatory, National Center for Atmospheric Research, courtesy of Joan Burkepile
Fig. 2.1	ESA/HIPPARCOS
Fig. 2.2	L. Golub
Fig. 2.3	J. M. Pasachoff
Fig. 2.4	W. Livingston, NOAO
Fig. 2.5	David Hathaway, NASA's Marshall Space Flight Center
Fig. 2.6	Rijks Museum, Amsterdam
Fig. 2.7	Jose M. Vaquero, Universidad de Extremadura
Fig. 2.8	Richard C. Willson, ACRIMsat/NASA with information from NASA and ESA satellites

Fig. 2.9 Solar Influences Data Analysis Center, Royal Observatory of Belgium
Fig. 3.1 J. M. Pasachoff
Fig. 3.2 J. M. Pasachoff
Fig. 3.3 J. M. Pasachoff
Fig. 3.4 Mt. Wilson Obs., Carnegie Institution for Science
Fig. 3.5 L. Golub
Fig. 3.6 NASA and Stanford, SDO/HMI
Fig. 3.7 W. Livingston, NOAO
Fig. 3.8 Swedish 1-m Solar Telescope (SST), Stockholm University and observer Vasco Henriques.
Fig. 3.9 Swedish 1-m Solar Telescope (SST), Stockholm University and observer Mats Löfdahl.
Fig. 3.10 SOHO/SOI-MDI
Fig. 3.11 J. M. Pasachoff
Fig. 3.12 Observatoire de Meudon
Fig. 3.13 Haleakala Observatory, Institute for Astronomy, U. Hawaii
Fig. 3.14 J. M. Pasachoff
Fig. 3.15 NASA/TRACE
Fig. 3.16 National Solar Observatory, Sacramento Peak
Fig. 3.17 J. M. Pasachoff
Fig. 3.18 from A. Title, *Selected Spectroheliograms*; observer R. W. Noyes
Fig. 3.19 J. M. Pasachoff
Fig. 3.20 J. M. Pasachoff
Fig. 3.21 Big Bear Solar Observatory, New Jersey Institute of Technology
Fig. 3.22 W. Livingston, National Solar Observatory
Fig. 3.23 NRL and NASA
Fig. 3.24 A. Daw, NASA
Fig. 4.1 L. Golub
Fig. 4.2 L. Golub
Fig. 4.3 Institute for Cosmic Ray Research, University of Tokyo
Fig. 4.4 SOHO/SOI-MDI
Fig. 5.1 L. Golub
Fig. 5.2 J. M. Pasachoff
Fig. 5.3 F. Espenak, GSFC
Fig. 5.4 J. M. Pasachoff
Fig. 5.5 A. B. Davis and J. M. Pasachoff, Williams College Eclipse Expedition, sponsored by the Committee for Research and Exploration of the National Geographic Society.
Fig. 5.6 J. M. Pasachoff
Fig. 5.7 High Altitude Observatory/NCAR
Fig. 5.8 Computer composites by Wendy Carlos based on originals from Jay Pasachoff, Jonathan Kern, William Wagner, Ronald Dantowitz, Nicholas Weber, Muzhou Lu, and Bryce Babcock – © 2013 and All Rights Reserved.

Fig. 5.9	F. Espenak, GSFC
Fig. 5.10	F. Espenak, GSFC
Fig. 5.11	SWAP from D. B. Seaton, ROB and ESA; eclipse from the Williams College Expedition/National Geographic Committee for Research and Exploration, computer analysis by Wendy Carlos; SOHO EIT from NASA's GSFC/ESA
Fig. 5.12	Institute for Astronomy, Haleakala Observatory
Fig. 5.13	L. Golub
Fig. 5.14	J. Raymond, SAO
Fig. 6.1	LASCO SOHO Consortium
Fig. 6.2	NASA/TRACE
Fig. 6.3	G. Nystrom, SAO
Fig. 6.4	NASA/TRACE
Fig. 6.5	G. Withbroe, NASA
Fig. 6.6	U.C., Berkeley
Fig. 6.7	Lockheed-Martin Missiles & Space, and Southwest Research Institute
Fig. 6.8	Johns Hopkins APL
Fig. 6.9	SAO and JAXA
Fig. 6.10	Lockheed-Martin and ISAS
Fig. 6.11	NASA, JHU/APL
Fig. 6.12a	NASA/MSFC/SAO, processed by E. Tavabi and S. Koutchmy
Fig. 6.12b	LMSAL/SAO/Montana State U./NASA
Fig. 7.1 – 7.8	L. Golub
Fig. 7.9	J. Lean, NRL
Fig. 8.1	Natural Resources Canada, Geological Survey of Canada
Fig. 8.2	L. Golub
Fig. 8.3	L. Golub
Fig. 8.4	L. Golub
Fig. 8.5	NASA/SDO/AIA
Fig. 8.6	LASCO SOHO Consortium
Fig. 8.7	L. Golub
Plate I	NASA, ESA and The Hubble Heritage Team (STScI/AURA)
Plate II	Global Oscillations Network Group
Plate III	National Solar Observatory, Kitt Peak
Plate IV	NASA/SDO/AIA
Plate V	NASA/TRACE
Plate VI	Jay Pasachoff, Ron Dantowitz, Nicholas Weber, and Muzhou Lu (Williams College/Clay Center Observatory/Dexter-Southfield School/NSF); NASA/SOHO/LASCO; NASA/SDO/AIA.
Plate VII	NASA/TRACE
Plate VIII	NASA/TRACE
Plate IXa,b	SOHO/SOI-MDI
Plate X	SOHO/SOI-MDI

Plate XI	JHU/APL
Plate XII	ISAS
Plate XIII	NOAA/POES
Plate XIV	NASA/SDO/AIA/GSFC
Plate XV	Jay M. Pasachoff (Williams College) and Ronald Dantowitz (Clay Center Observatory/Dexter-Southfield School)/National Geographic Society Committee for Research and Exploration
Plate XVI	ISS Expedition 23 Crew, ISAL: NASA's Instrument Synthesis & Analysis Laboratory

Bibliography

The following are books, articles and reports, at varying technical levels, that we suggest to readers interested in studying in more detail some of the subjects covered in this book. We have generally chosen material that is accessible to readers without scientific training, but we have also included references to some material that will be of interest to the more technically minded reader. The two categories are separated below.

For the general reader

- Barrow, J. D., and Tipler, F. J., *The Anthropic Cosmological Principle* (Oxford University Press, Oxford, UK, 1988).
- Beatty, J. K., Peterson, Collins, Carolyn, and Chaikin, Andrew, eds., *The New Solar System*, 4th edition (Cambridge University Press, 2000).
- Bester, Alfred, *The Life and Death of a Satellite* (Little Brown, Boston, 1966).
- Brekke, Pål, *Our Explosive Sun: A Visual Feast of Our Source of Light and Life* (Springer, Heidelberg, 2012).
- Chapman, Allen, "The Transits of Venus." *Endeavour*, vol. 22(4), pp. 148–151, 1998.
- Chapman, William J., *Music of the Sun: The Story of Helioseismology* (Oxford, Oneworld, 2006).
- Cohen, Richard, *Chasing the Sun: The Epic Story of the Star That Gives Us Life* (Random House, 2011).
- Eddy, John A., *The Sun, The Earth, and Near-Earth Space: A Guide to the Sun-Earth System* (NASA, 2009), bookstore.gpo.gov.
- Fernie, J. D., "Transit, Travels and Tribulations, I." American Scientist, volume 85: 120–122, 1997.

- — "Transit, Travels and Tribulations, II." American Scientist, volume 85: 418–421, September-October 1997.
- — "Transit, Travels and Tribulations, III." American Scientist, volume 86: 123–126, March-April 1998.
- — "Transit, Travels and Tribulations, IV." American Scientist, volume 86: 422–425, September-October 1998.
- Grice, Noreen, *Touch the Sun* (Joseph Henry Press, Washington, DC, 2005). In Braille with figures embossed so that the visually impaired can feel the shapes, www.youcandoastronomy.com
- Guillermier, Pierre, and Koutchmy, Serge, *Total Eclipses: Science, Observations, Myths and Legends* (Chichester: Praxis and Springer, 1999); translated from *Eclipses Totales: Histoire, Découvertes, Observations* (Paris: Masson, 1998).
- Harrington, Philip S., *Eclipse! The What, Where, When, Why & How Guide to Watching Solar and Lunar Eclipses* (New York/Chichester, Wiley, 1997).
- Kragh, Helge, "The Solar Element: A Reconsideration of Helium's Early History," *Annals of Science*, 66:2, 157–182 (2009).
- Hirsh, Richard F., *Glimpsing an Invisible Universe: The Emergence of X-Ray Astronomy* (Cambridge University Press, 1985).
- Hughes, Stefan, *Catchers of the Light: The Forgotten Lives of the Men and Women who First Photographed the Heavens* (ArtDeCiel, Cyprus, 2014); pdf and e-book at catchersofthelight.com
- Imbrie, John, and Imbrie, Katherine Palmer, *Ice Ages: Solving the Mystery*, 2nd. ed. (Cambridge: Harvard University Press, 1986).
- IPCC Fourth Assessment Report, *Climate Change 2007: The Physical Science Basis*. Intergovernmental Panel on Climate Change.
- Jenkins, Jamey L., *Observing the Sun: A Pocket Field Guide* (Springer, New York, 2013).
- Lang, Kenneth R., *Sun, Earth & Sky* (Springer, New York, 2006).
- — *The Sun From Space*, Astronomy and Astrophysics Library, Springer 2009.
- Littmann, Mark, Espenak, Fred, and Willcox, Ken, *Totality: Eclipses of the Sun* (Oxford University Press, New York, 2008).
- Lomb, Nick, *Transits of Venus* (The Experiment Publishing, New York, 2012).
- Menzel, Donald H., *Our Sun*, 2nd edition (Harvard University Press, 1959).
- Mitton, Simon, *Fred Hoyle: A Life in Science* (Cambridge University Press, 2011).
- Nath, Biman, *The Story of Helium and the Birth of Astrophysics* (Heidelberg, Springer, 2012).

- Noyes, Robert W., *The Sun, Our Star* (Harvard University Press, 1982).
- Pasachoff, Jay M., *Peterson Field Guide to the Stars and Planets*, 4th ed. (Houghton Mifflin Harcourt, Boston, 2000, 2012 printing).
- Pasachoff, Jay M., *Peterson First Guide to Astronomy*, 2nd ed. (Houghton Mifflin Harcourt, Boston, 2014).
- Pasachoff, Jay M., and Filippenko, Alex, *The Cosmos: Astronomy in the New Millennium*, 4th edition (Cambridge University Press, 2014).
- Pasachoff, Jay M., and MacRobert, Alan, "Is the Sunspot Cycle About to Stop?" *Sky & Telescope*, Sept., **122**, #3, 12–13, 2011; http://www.skyandtelescope.com/community/skyblog/newsblog/123844859.html.
- Pasachoff, Jay M., and Suer, Terry-Ann, "The Origin and Diffusion of the H and K Notation," *Journal of Astronomical History and Heritage*, 13 (2), 121–127, 2010; http://adsabs.harvard.edu/abs/2010JAHH...13..120P
- Phillips, Kenneth J. H., *Guide to the Sun* (Cambridge University Press, 1992).
- Sheehan, William, and Westfall, John, *Celestial Shadows Eclipses, Transits and Occultations* (Springer, New York, 2014).
- Stevens, William K., *The Change in the Weather: People, Weather and the Science of Climate* (Delacorte Press, 2001).
- Ward, Peter D., and Brownlee, Donald, *Rare Earth: Why Complex Life is Uncommon in the Universe* (Copernicus Books, 2001).
- Wulf, Andrea, *Chasing Venus: The Race to Measure the Heavens* (Alfred A. Knopf, New York, 2012).
- Zirker, Jack B., *The Magnetic Universe: The Elusive Traces of an Invisible Force* (Johns Hopkins University Press, Baltimore, 2009).

More technical works

- Ashchwanden, Markus, *Physics of the Solar Corona: An Introduction with Problems and Solutions* (Springer, New York, 2009); http://www.lmsal.com/~aschwand/bookmarks_books.html
- Bahcall, John N., *Neutrino Astrophysics* (Cambridge University Press, 1989).
- Bhatnagar, Arvind, and Livingston, William, *Fundamentals of Solar Astronomy* (World Scientific, Hackensack, NJ, 2005).
- Cirtain, Jonathan W., et al., "Energy release in the solar corona from spatially resolved magnetic braids." *Nature*, **493**, 501, 2013.
- Eddy, John A., "The Maunder Minimum," *Science* **192**, 1189, 1976.
- Foukal, Peter V., *Solar Astrophysics*, 3rd ed. (Wiley, 2013).

- Golub, Leon, and Pasachoff, Jay M., *The Solar Corona*, 2nd ed. (Cambridge University Press, 2010).
- Gough, Douglas O., Leibacher, John W., Scherrer, Philip H., and Toomre, Juri, "Perspectives in Helioseismology." *Science* **272**, 1281–1283, 1996.
- Gray, L. J., et al., "Solar Influences on Climate," *Reviews of Geophysics* **48**, 4, RG4001, 2010.
- Jäger, Jill, *Climate and Energy Systems* (John Wiley and Sons, New York, 1983).
- Kaneyuki, Kenji, and Scholberg, Kate, "Neutrino Oscillations," *American Scientist*, May-June 1999, pp. 222–231.
- Lamb, Hubert H., *Climate: Past, Present and Future*, Vol. 1, Routledge, London 1972, 2001, 2006).
- Lamb, Hubert H., *History and the Modern World*, 2nd ed. (Routledge, London, 1982, 2005).
- Lean, Judith, "Variations in the Sun's Radiative Output," *Reviews of Geophysics*, **29**, 4, 505, 1991.
- Lean, Judith, "The Sun's Variable Radiation and Its Relevance for Earth," *Annual Reviews of Astronomy & Astrophysics*, **35**, 33, 1997.
- Maor, Eli, *June 8, 2004–Venus in Transit* (Princeton University Press, 2000).
- Mariska, John T., *The Solar Transition Region* (Cambridge University Press, 1992).
- National Research Council: 1994, *Solar Influences on Global Change*, National Academy Press, Washington, D.C.
- Newton, R. R., *Medieval Chronicles and the Rotation of the Earth* (The Johns Hopkins University Press, 1972).
- Sagan, Carl, and Mullen, George, "Earth and Mars: Evolution of atmospheres and surface temperatures," *Science*, **177**, 52, 1972.
- Sonett, Charles P., Giampapa, Mark S., and Matthews, Mildred S., *The Sun In Time* (The University of Arizona Press, Tucson, 1991).
- Stephenson, F. Richard, *Historical Eclipses and Earth's Rotation* (Cambridge University Press, 1997).
- Suess, Steven T., and Tsurutani, Bruce T., eds., *From the Sun: Auroras, Magnetic Storms, Solar Flares, Cosmic Rays* (American Geophysical Union, Washington, DC, 1998).
- Ulrich, Roger K., "Solar neutrinos and variations in the solar luminosity," *Science*, **190**, 619, 1975.
- Wenzel, Donat, *The Restless Sun* (Smithsonian Institution Press, 1989)
- Williams, Jack, *The Weather Book*, 2nd. ed. (Vintage Books, New York, 1997).

Index